P9-AQN-516

Applications of Percolation Theory

Applications of Percolation Theory

Muhammad Sahimi

University of Southern California

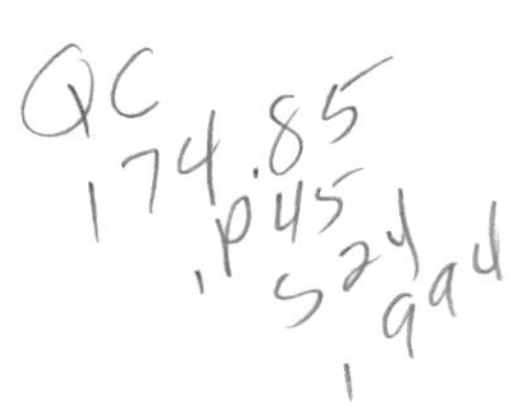

UK Taylor & Francis Ltd, 4 John St., London WC1N 2ET

USA Taylor & Francis Inc., 1900 Frost Road, Suite 101, Bristol PA 19007

British Library Cataloguing in Publication Data

A catalogue record for this book is available from the British Library

ISBN 0 7484 0075 3 (cloth)
0 7484 0076 1 (paper)

Cover design by John Leath

Typeset in Great Britain by Keyword Publishing Services
Printed in Great Britain by Burgess Science Press, Basingstoke on paper which has a specified pH value on final paper manufacture of not less than 7.5 and is therefore 'acid free'.

Dedicated to

my native country, Iran,
may she live in peace, freedom, and prosperity;
and to
the memory of my young brother, Ali,
who lost his life for her.

Contents

Preface

Disorder plays a fundamental role in many processes of industrial and scientific interest. Of all the disordered systems, porous media are perhaps the best known example, but other types of disordered systems such as polymers and composite materials are also important and have been studied for a long time. With the advent of new experimental techniques, it has become possible to study the structure of such systems and gain a much deeper understanding of their properties. New techniques have also allowed us to design the structure of many disordered systems in such a way that they possess the properties that we desire.

During the past two decades, the development of a class of powerful theoretical methods has enabled us to interpret our experimental observations, and predict many properties of disordered systems. Included in this class are renormalization group theory, modern versions of the effective medium approximation, and percolation theory. Concepts of percolation theory have, however, played one of the most important roles in our current understanding of disordered systems and their properties. This book attempts to summarize and discuss some of the most important applications of percolation theory to modeling of various phenomena in disordered systems. Among such applications porous media problems have perhaps received the widest attention. Thus, we now have a fairly good understanding of two-phase flow problems in porous media, and recognize that, at least in certain limits, such problems represent percolation phenomena. Oil recovery processes have thus benefited from the application of percolation to two-phase flow problems. Reaction and diffusion in porous materials, such as catalysts and coal particles, have also benefited from the insight that percolation has provided us. Such applications are discussed in detail in this book. However, other well-known applications of percolation, such as those to polymers and gels, composites materials, and rock masses are also discussed.

In this book we consider applications of percolation theory to those phenomena for which there are well-defined percolation models, *and* a direct comparison between the predictions of the models and experimental data is possible. Percolation has been applied to many phenomena; however, it has not always been possible to compare the predictions with the data, and therefore it has not always been possible to check the quantitative accuracy of the predictions. Although a theoretician may justifiably argue that, "an application is an application is an application," the scope of the book is limited to such "practical" applications, since it is impossible to discuss in one book all applications of percolation theory, and this is a book written by an application oriented chemical engineer. New applications of percolation are still being developed, and in the coming years such applications will find widespread use in many branches of science and technology.

Over the past decade Dietrich Stauffer has greatly contributed to my understanding of percolation theory, disordered systems, and critical phenomena. He has done this through his "referee's reports", e-mail messages, letters, and our collaborations on various problems. Without his constant encouragement and support this book would not have been written. He also read most of the book and offered constructive criticism and very useful suggestions. I am deeply grateful to him. I would also like to thank Ted Davis and Skip Scriven who introduced me to percolation, and Barry Hughes for his many stimulating discussions and fruitful collaboration. Many other people have contributed to my understanding of percolation theory, a list of whom is too long to be given here. I would like to thank all of them.

Most of this book was written while I was visiting the HLRZ Supercomputer Center at KFA Jülich, Germany, as an Alexander von Humboldt Foundation Research Fellow. I would like to thank Hans Herrmann and the Center for their warm hospitality, and the Foundation for financial support.

Muhammad Sahimi
Los Angeles,
April 1993

1
Connectivity as the essential physics of disordered systems

1.0 Introduction

It is a fact of life, which is as challenging to the mind of the scientist as it is frustrating to his or her aspirations, that *nature is disordered*. In nowhere but the theoretician's supermarket can we buy clean, pure, perfectly characterized and geometrically immaculate systems. An engineer works in a world of composites and mixtures (how much more so the biologist). Even the experimentalist who focusses on the purest of subtances, exemplified by carefully grown crystals, can seldom escape the effects of defects, trace impurities, and finite boundaries. There are few concepts in science more elegant to contemplate than an infinite, perfectly periodic crystal lattice, and few systems as remote from experimental reality. We are therefore obliged to come to terms with *disordered structures*; variation in shape and constitution often so ill-characterized that we must deem it to be random if we are to describe it – apparent randomness in the morphology of the system. The morphology of a system has two major aspects: *topology*, the interconnectiveness of individual microscopic elements of the system; and *geometry*, the shape and size of these individual elements.

As if this were not bad news enough, we know that however random the stage upon which the drama of nature is played out, it is also at times very difficult to follow the script. We believe, at least above the quantum mechanical level, in the doctrine of determinism, yet important *continuum* systems exist in which deterministic descriptions are beyond hope. Typical examples are diffusion and Brownian motion where, over certain length scales, we observe an apparent random process, or *disordered dynamics*.

Nature, then, is disordered both in her structure and the processes she supports. Indeed the two types of disorder are often concurrent and coupled. An example, dear to the heart of the author, is fluid flow through a porous medium where the interplay between the disordered structure of the pore space and the dynamics of fluid motion gives rise to a rich variety of phenomena, some of which will be discussed in this book.

Despite the rather obvious randomness in nature, these topics were, for several decades, familiar to most physicists, engineers, and others only in the form of statistical mechanics, or in the application of such equations as the Boltzmann equation. Research in these fields made remarkable progress by taking advantage of periodic structures. However, as one always has to confront the real world, it became apparent that a statistical physics of disordered systems must be devised to provide methods for deriving macroscopic properties of such systems from laws governing the microscopic world, or alternatively for deducing microscopic properties of such systems from the macroscopic information that can be observed by experimental techniques. Such a statistical physics of disordered systems should take into account the effect of *both* morphology and geometry of the systems. While the role of geometry was already appreciated in the early years of this century, the effect of topology was ignored for many decades, or was treated in an unrealistic manner, simply because it was thought to be too difficult to be taken into account.

A study of the history of science also shows that progress in any research field is not usually made with a constant rate, but rather in a sporadic manner. There are periods when a problem looks so difficult that we do not even know where to start, and periods when some epoch-making discoveries or inventions remove an obstacle to progress and thus enable a great advance. An example is the discovery of a new class of superconducting materials in 1986. Bednorz and Muller (1986) showed that it is possible to have superconductivity in (La, Ba)CuO alloys at temperatures $T_c > 30$K. Subsequently it was shown by Tagaki *et al.* (1987) that the phase $La_{2-x}Ba_xCuO_4$ with $x \sim 0.15$ is responsible for bulk superconductivity with $T_c \sim 35$K. Since then hundreds, and perhaps thousands, of research papers have been written on the subject of high-temperature superconductivity, and the literature on the subject has become very forbidding. The discovery had such an impact on the subject that Bednorz and Muller were honoured with a Noble Prize in physics in 1989. Over the past two decades, statistical physics of disordered systems has been in this rapidly moving stage of progress, partly because standard methods for calculating the average properties of disordered systems have been established from the theoretical side, and also more and more experimental results have been accumulated thanks to many novel experimental techniques. But perhaps the most important reason for the rapid development of the statistical physics of disordered systems is that the role of interconnectivity of the microscopic elements of a disordered system and its effect on the macroscopic properties of the system has been appreciated. This has been possible through the development and application of percolation theory, the subject of this book.

However, this is an applications oriented book written by a chemical engineer, not a book on mathematical or computational methods of a theoretical physicist. As such, the book is necessarily biased.

1.1 What is percolation?

When I was a graduate student, I was living in Minneapolis, Minnesota, and attending the University of Minnesota. Minneapolis is a wonderful city of many lakes and naturally I tried to live near a lake to enjoy the beautiful summer, and to watch little kids skate on the lake during the long winter. The magnificent Mississippi river runs through Minneapolis, dividing the campus of the university into east and west bank sections, and the chemical engineering department where I was a graduate student was on the east bank. I was living in southeast Minneapolis near Lake Calhoun, about eight miles from the campus. Every winter we had many snow storms that would cause temporary closure of many streets in one or both directions for sweeping the snow. Even during the long summer some streets would be closed for repairing the damage caused by the long winter. There are many other types of defects or disorder in the structure of the streets of Minneapolis. There is a big K-Mart store that blocks Nicolete Avenue, one of the most important routes in the city that starts in the downtown area and ends in the suburb in the south. Many railways also cut the streets, and many lakes have created natural blockage for many streets. It often seemed as if the streets were closed at random! Now suppose that we idealize the streets of Minneapolis as the bonds of a very large square network, and block at random some of the streets by a heavy snow, a rail track, a lake, or a store like K-Mart (see Fig. 1.1). As a PhD student of two of the most

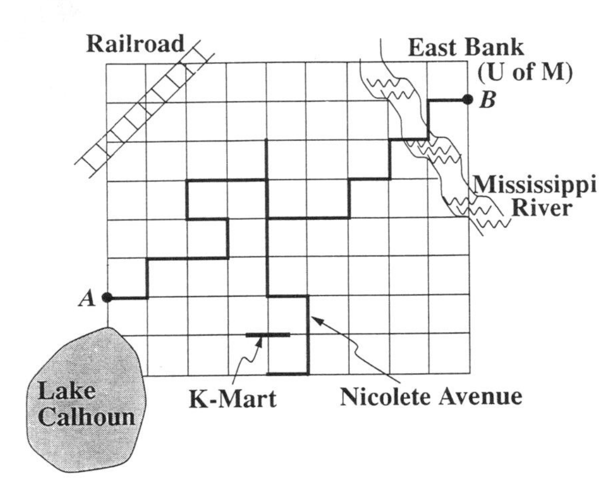

Figure 1.1 *Idealization of streets of Minneapolis.*

accomplished and famous chemical engineering professors in the country in one of the top-ranked departments, I was expected to work on my research topics six days a week, from Monday through Saturday, and many Sundays. I would leave my apartment to go to the department and wonder what fraction of streets of Minneapolis would have to be open to traffic in order for me to reach the department from my apartment. Obviously, if too many streets were closed, I could not reach the department, and therefore would not get anything done. On the other hand, if most streets were open, almost any route would take me to the department. Therefore, there must be a critical value of the fraction of open streets below which I would never reach the department, but above which I could always get to work.

Let us consider another example. Suppose that, instead of representing the streets of Minneapolis, the bonds of the square network of Fig. 1.1 represent resistors. We assign a unit resistance to the open bonds, and an infinite resistance to the closed ones (i.e., the closed bonds are insulators). Suppose also that we impose a unit voltage at point A in Fig. 1.1, and a zero voltage at B. What fraction of the bonds have to have a finite resistance in order for electrical current to flow from A to B? This is an important question because its answer tells us what (volume) fraction of a disordered material, such as carbon black composites that are routinely used in many applications, has to be conducting in order for the composite as a whole to be conducting. The reader can probably see the similarity between the flow of current in the composite and traffic flow in the streets of Minneapolis. As in the first example, if too many bonds are insulating, no macroscopic current will flow from A to B, whereas for a sufficiently large number of resistors one can have electrical current between the two points.

Consider a third example. Imagine that the bonds of the network of Fig. 1.1 are pores of a porous medium, e.g., an underground oil reservoir. In reality, no porous medium looks as regular as the square network, but as an idealization this network serves a useful purpose. Now suppose that the pores are filled with oil, and that there are two wells in the system, one at A and the other at B. We would like to push oil out of this porous medium by injecting water into the system at A (the injection well) and producing oil at B (the production well). Oil and water do not mix with each other, and therefore we assume that each pore is filled with either oil or water. As the water is injected and pushed into the reservoir, it tries to find the *smallest pores* that it can reach and expel the oil from it (why this is so is explained in Chapter 7). In reality, the process is more complex than this, but we ignore all the complications. The expelled oil is produced at well B. What fraction of the pores are filled with water when it reaches the production well at B for the first time (this is called *breakthrough point*)? In other words, we would like to know what fraction of the pores lose their oil, and how much oil is produced at well B at the breakthrough point. This is obviously an important question. In the early 1980s when the price of oil was as high as $40/barrel and the United States suffered a traumatic energy crisis, intense research was done to answer this and related questions. Even with the relatively cheap oil currently available, intense research is still going on.

Questions such as these are answered by percolation theory. Percolation tells us when a system is *macroscopically* open to a given phenomenon. For example, it can tell us when one can have flow of traffic from one side of a town to its opposite, when electrical current can flow from one side of a composite to the opposite, and how much oil one can extract from an oil reservoir. The point at which the percolation transition between an open system and a closed one takes place for the first time is the *percolation threshold* of the system, and the behavior of the system close to this point is of prime interest and importance. Because a percolation network is created by simply blocking bonds at random, percolation is also useful as a simple model of disordered systems. Moreover, since the main concepts of percolation theory are very simple, writing a computer program for simulating a percolation process is straightforward, and so percolation can also serve as a simple tool for introducing students to computer simulations. Deutscher *et al.* (1983), Bunde and Havlin (1991), Stauffer and Aharony (1992), and Hughes (1993) have emphasized the foundations of percolation theory. This book attempts to summarize some important applications of percolation.

1.2 The scope of the book

Over the past two decades percolation has been applied to modeling a wide variety of phenomena in disordered systems. It is impossible to discuss all such applications in one book. In selecting those applications that are discussed in this book, three criteria were used.

(i) The application is quantitative, in the sense that there is a quantitative agreement between the predictions of percolation and experimental data.
(ii) The problem is of broad interest, or has broad industrial, technological, or scientific importance.
(iii) This author has a clear understanding of the problem and how the application of percolation is made.

Based on these criteria, I selected twelve classes of problems to which percolation has been applied. In discussing each case, it is assumed that the reader already knows the essence of the problem. Stauffer and Aharony (1992) provide an excellent and simple introduction to the percolation concepts. Therefore, Chapter 2 contains only a summary of the main properties that will be used in the rest of this book. Every effort is made to explain the percolation approach in simple terms. In all cases, the predictions are compared with the experimental data to establish the relevance of percolation concepts and their application to the problem. We also give what we believe are the most relevant references to each subject, or provide the reference to a recent review on the subject.

We finish this chapter by mentioning a few other applications of percolation which, although they do not meet at least one of the above three criteria, represent important and useful applications of percolation theory. The literature cited below and in the subsequent chapters is not necessarily the first or the best work on the subject. It represents what was known to me at the time of writing this book, or what I considered to be the most relevant.

1.3 Applications of percolation that are not discussed in this book

Now that we know what kind of applications of percolation are discussed in this book, let us mention a few other important applications and references to them that will not be discussed here.

1.3.1 Percolation model of galactic structures

It has been shown that the process of star formation may be a percolation process, and that the percolation transition from a disconnected to a connected state plays an important role in the stabilization and control of star formation. For example, for galaxies whose formation can be described by a percolation process, the main feature, namely, the presence of spiral arms, is a consequence of the proximity to the percolation transition. The interested reader should consult Seiden and Schulman (1990) for a review of this interesting application of percolation.

1.3.2 Multicomponent percolation

In all cases that are discussed in this book, the percolation system is essentially a binary mixture of two components. For example, in the above examples one has a two component mixture of open and closed streets, resistors and insulators, and pores filled with oil and water. However, there are many cases in which the system may consist of three or more components. For example, charge transfer salts are random mixtures of a metal M_1 with another metal or element M_2 in which M_1 and M_2 react and produce $M_1^+M_2^-$, if M_1 and M_2 are in close proximity. The reaction is reversible, and can reproduce M_1 and M_2 again. Examples are $Cs_{1-x}Te_x$, $Cs_{1-x}Sb_x$, $Na_{1-x}Sn_x$, and many others. One is interested in the conductivity of this system, which can be treated by a three-component (M_1, M_2, and $M_1^+M_2^-$) percolation model. This and many other applications of multicomponent percolation are discussed by Halley (1983).

1.3.3 Glass transition

It may not be obvious to the reader how the transition between a liquid and glassy state may have anything to do with percolation, since percolation in its fundamental form is a static process, whereas the liquid–glass transition is the result of the system moving away from complete metastable equilibrium. However, it has been shown that percolation can help us understand certain aspects of liquid–glass transition, although this application of percolation remains controversial. Earlier reviews were given by Grest and Cohen (1983) and Zallen (1983). For the most recent reference on the subject see Wittmann *et al.* (1992).

1.3.4 The behavior of supercooled water

A snapshot of liquid water shows that this system, which is a hydrogen-bonded liquid, has the structure of a network above its percolation threshold. However, the network is well above its percolation threshold, and therefore the classic random percolation discussed in the above examples is not applicable to this problem. Instead, a correlated percolation model has been found to explain many unusual properties of supercooled water. For the most recent references on the subject see Stanley *et al.* (1983), Blumberg *et al.* (1984), and Sciortino *et al.* (1990).

1.3.5 Dynamic percolation

In the classical percolation problem the configuration of the system does not change with time. That is, the probability that a bond is open or closed is independent of time. In many situations this description of a disordered system is inadequate since the configuration of the system changes with time. For example, in polymeric ionic conductors above the glass transition temperature, the electrical conductivity is dominated by microscopic motion of the medium. Another example is the problem of oxygen binding to haemoglobin and myoglobin. It is known that the entrance to the haeme pocket is blocked by a number of side chains and thus oxygen could not bind if the side chains were fixed at their equilibrium positions. Because of the dynamic nature of proteins one may expect that a gate would fluctuate between open and closed positions. Thus one has to deal with a diffusion process in a disordered medium whose structure varies with time.

To deal with such problems, various dynamic percolation models in which the probability that a bond or site of a network is open varies with time have been proposed by Druger *et al.* (1985), Harrison and Zwanzig (1985), Sahimi (1986), and Bunde *et al.* (1991); see also Granek and Nitzan (1989, 1990), and Loring (1991). Another type of dynamic percolation was developed by Bunde and coworkers for solid ionic conductors; see Bunde *et al.* (1986a, b), Harder *et al.* (1986), and Roman *et al.* (1986). Finally, quite different dynamic percolation models have been suggested for gelation phenomena, and microemulsion systems. They are discussed in Chapters 11 and 12.

References

Bednorz, J.G. and Muller, K.A., 1986, *Z. Phys. B* **64**, 189.
Bunde, A., Dieterich, W., and Roman, H.E., 1986a, *Solid State Ionics* **119**, 747.
Bunde, A., Ingram, M.D., Maass, P., and Ngai, K.L., 1991, *J. Phys. A* **24**, L881.
Bunde, A., Harder, H., and Dieterich, W., 1986b, *Solid State Ionics* **119**, 358.
Bunde, A. and Havlin, S. (eds.), 1991, *Fractals and Disordered Systems*, Berlin, Springer.
Blumberg, R.L., Stanley, H.E., Geiger, A., and Mausbach, P., 1984, *J. Chem. Phys.* **80**, 5230.
Deutscher, G., Zallen, R., and Adler, J. (eds.), 1983, *Percolation Structures and Processes*, Bristol, Adam Hilger.
Druger, S.D., Ratner, M.A., and Nitzan, A., 1985, *Phys. Rev. B* **31**, 3939.
Granek, R. and Nitzan, A., 1989, *J. Chem. Phys.* **90**, 3784.
Granek, R. and Nitzan, A., 1990, *J. Chem. Phys.* **92**, 1329.
Grest, G.S. and Cohen, M.H., 1983, in *Percolation Structures and Processes*, edited by G. Deutscher, R. Zallen and J. Adler, Bristol, Adam Hilger, p. 187.
Halley, J.W., 1983, in *Percolation Structures and Processes*, edited by G. Deutscher, R. Zallen and J. Adler, Bristol, Adam Hilger, p. 323.
Harder, H., Bunde, A., and Dieterich, W., 1986, *J. Chem. Phys.* **85**, 4123.
Harrison, A.K. and Zwanzig, R., 1985, *Phys. Rev. A* **32**, 1072.
Hughes, B.D., 1993, *Random Environments and Random Walks*, London, Oxford University Press.
Loring, R.F., 1991, *J. Chem. Phys.* **94**, 1505.
Roman, H.E., Bunde, A., and Dieterich, W., 1986, *Phys. Rev. B* **33**, 3439.
Sahimi, M., 1986, *J. Phys. C* **19**, 1311.
Sciortino, F., Poole, P.H., Stanley, H.E., and Havlin, S., 1990, *Phys. Rev. Lett.* **64**, 1686.
Seiden, P.E. and Schulman, L.S., 1990, *Adv. Phys.* **39**, 1.
Stanley, H.E., Blumberg, R.L., and Geiger, A., 1983, *Phys. Rev. B* **28**, 1626.
Stauffer, D. and Aharony, A., 1992, *Introduction to Percolation Theory*, 2nd ed., London, Taylor & Francis.
Tagaki, H., Uchida, S., Kitazawa, K., and Tanaka, S., 1987, *Jap. J. Appl. Phys. Lett.* **26**, L123.
Wittmann, H.-P., Kremer, K., and Binder, K., 1992, *J. Chem. Phys.* **96**, 6291.
Zallen, R., 1983, *The Physics of Amorphous Solids*, New York, Wiley.

2
Elements of percolation theory

2.0 Introduction

Percolation processes were first developed by Flory (1941) and Stockmayer (1943) to describe how small branching molecules react and form very large macromolecules. This polymerization process may lead to *gelation*, i.e., to the formation of a very large network of molecules connected by chemical bonds, the key concept of percolation theory. However, Flory and Stockmayer developed their theory of gelation for a special kind of network, namely, the Bethe lattice, an endlessly branching structure without any closed loops (see Fig. 2.1). We return to the Flory–Stockmayer theory in Chapter 11.

In the mathematical literature, percolation was introduced by Broadbent and Hammersley (1957). They originally dealt with the concept of the spread of hypothetical fluid particles through a random medium. The terms fluid and medium were viewed as totally general: a fluid can be liquid, vapor, heat flux, electric current, infection, a solar system, and so on. The medium – where the fluid is carried – can be the pore space of rock, an array of trees, or the universe. Generally speaking, the spread of a fluid through a disordered medium involves some random elements, but the underlying mechanism(s) for this might be one of two very different types. In one type, the randomness is ascribed to the *fluid*: the fluid particles decide where to go in the medium. This is the familiar *diffusion process*. In the other type, the randomness is ascribed to the *medium*: the medium dictates the paths of the particles. This was the new situation that was considered by Broadbent and Hammersley (1957). Hence it also demanded its own terminology. It was decided to name it a *percolation process*, since it was thought that the spread of the fluid through the random medium resembled the flow of coffee in a percolator.

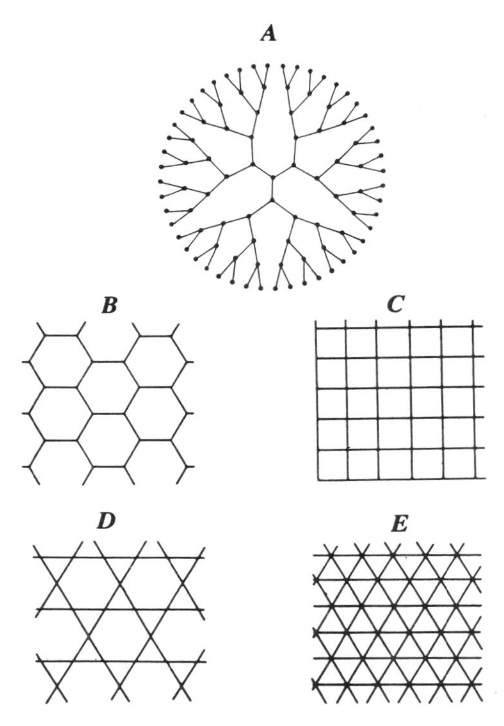

Figure 2.1 *Some regular lattices. (A) The Bethe lattice ($Z = 3$). (B) The honeycomb lattice ($Z = 3$). (C) The square lattice ($Z = 4$). (D) Kagomé lattice ($Z = 4$). (E) The triangular lattice ($Z = 6$).*

2.1 Definitions and percolation thresholds

We first discuss percolation processes on regular networks, and then discuss them for random networks and continua. The classical percolation theory centers around two problems. In the *bond percolation problem*, the bonds of the network are either *occupied* (i.e., they are open to flow, diffusion and reaction, they are microscopic conducting elements of a composite, etc.), randomly and independently of each other with probability p, or are *vacant* (i.e., they are closed to flow or current, or have been plugged, they are insulating elements of a composite, etc.) with probability $1 - p$. For a large network, this assignment is equivalent to removing a fraction $1 - p$ of all bonds at random. Two sites are called *connected* if there exists at least one path between them consisting solely of occupied bonds. A set of connected sites bounded by vacant bonds is called a *cluster*. If the network is of very large extent and if p is sufficiently small, the size of any connected cluster is small. But if p is close to 1, the network should be entirely connected, apart from occasional small holes. At some well-defined value of p, there is a transition in the topological structure of the random network from a macroscopically disconnected structure to a connected one; this value is called the *bond percolation threshold*, p_{cb}. This is the *largest* fraction of occupied bonds below which there is no sample-spanning cluster of occupied bonds.

Similarly, in the *site percolation problem* sites of the network are occupied with probability p and vacant with probability $1 - p$. Two nearest-neighbor sites are called connected if they are both occupied, and connected clusters on the network are again defined in the obvious way. There is a site percolation threshold p_{cs} above which an infinite (sample-spanning) cluster of occupied sites spans the network. We should point out that, depending on a specific application, many variants of bond or site percolation have been developed, some of which will be discussed in this book.

The derivation of the exact values of p_{cb} and p_{cs} has been possible to date only for certain lattices related to the Bethe lattice and for a few two-dimensional lattices. The percolation thresholds of three-dimensional networks have been calculated numerically by Monte Carlo simulations or other techniques. For the Bethe lattice it can be shown that (Fisher and Essam 1961)

$$p_{cb} = p_{cs} = \frac{1}{Z - 1}, \tag{2.1}$$

where Z is the coordination number of the lattice, i.e., the number of bonds connected to the same site. Figure 2.1 shows some regular two-dimensional lattices. We compile the current estimates of p_{cb} and p_{cs} for common two- and three-dimensional lattices in Tables 2.1 and 2.2 respectively. In general, $p_{cb} \leqslant p_{cs}$. These tables show that the product $B_c = Zp_{cb}$ is essentially an invariant of percolation networks. For a d-dimensional system, $B_c \simeq d/(d - 1)$. The significance of B_c is discussed below.

Table 2.1 *Currently accepted values of the percolation thresholds of some two-dimensional networks*

Network	Z	p_{cb}	$B_c = Zp_{cb}$	p_{cs}
Honeycomb	3	$1 - 2\sin(\pi/18) \simeq 0.6527$*	1.96	0.6962
Square	4	1/2*	2	0.5927
Kagomé	4	0.522	2.088	0.652
Triangular	6	$2\sin(\pi/18) \simeq 0.3473$*	2.084	1/2*

*Exact result.

Table 2.2 *Currently accepted values of the percolation thresholds of some three-dimensional networks*

Network	Z	p_{cb}	$B_c = Zp_{cb}$	p_{cs}
Diamond	4	0.3886	1.55	0.4299
Simple cubic	6	0.2488	1.49	0.3116
BCC	8	0.1795	1.44	0.2464
FCC	12	0.198	1.43	0.119

2.2 Computer generation of percolation clusters

Generating a percolating lattice by randomly removing sites or bonds, although easy, is often not suitable for practical applications because, in addition to the sample-spanning cluster, this method also generates isolated finite clusters. In most applications one works only with the sample-spanning cluster (or the process of interest starts with a single cluster), and therefore we must first delete all isolated clusters from the system, which is time-consuming. Alternatively, we can use a method, developed by Leath (1976) and Alexandrowicz (1980), that generates only the sample-spanning (or the largest) cluster. In this method one starts with a single occupied site at the center of the lattice, identifies its nearest-neighbor sites and considers them occupied and adds them to the cluster if random numbers $0 < r < 1$, attributed to the neighbors, are less than the fixed value p. The perimeter sites, the nearest-neighbor empty sites of the occupied sites, are found and the process of occupying such sites continues in the same manner. If a selected perimeter site is not occupied, then it remains unoccupied forever. The generalization of this method for generating a cluster of occupied bonds is obvious.

An important task in computer simulations of percolating systems is to count the number of clusters of a given size, or the fraction of occupied sites or bonds that belong to the sample-spanning cluster. For example, during displacement of a fluid A by another immiscible fluid B in a porous medium we may need to know the number of islands or blobs of fluid A of a given size that are completely surrounded by B, which is equivalent to knowing the number of clusters of a given size within the context of a percolation model of fluid displacement in a porous medium (see Chapter 7). An efficient algorithm for doing this was developed by Hoshen and Kopelman (1976). A computer program implementing their method is given by Stauffer and Aharony (1992).

2.3 Percolation quantities

In addition to the percolation thresholds, the topological properties of percolation networks are characterized by several important quantities.

(i) *Percolation probability* $P(p)$. This is the probability that, when the fraction of occupied bonds is p, a given site belongs to the infinite (sample-spanning) cluster of occupied bonds.

(ii) *Accessible fraction* $X^A(p)$. This is that fraction of occupied bonds belonging to the infinite cluster.

(iii) *Backbone fraction* $X^B(p)$. This is the fraction of occupied bonds in the infinite cluster which actually carry flow or current, since some of the bonds in the cluster are dead-end and do not carry any flow. The backbone of a percolating system plays a fundamental role in its transport properties, because the tortuosity of the transport paths is controlled by the structure of the backbone. We shall discuss the structure of the backbone shortly.

(iv) *Correlation length* $\xi_p(p)$. This is the typical radius of the connected clusters for $p < p_c$, and the length scale over which the random network is macroscopically homogeneous (i.e., the length scale over which the properties of the system are independent of its linear size L) for $p > p_c$. Thus, in any Monte Carlo simulations of percolation we must have $L >> \xi_p$ for the results to be independent of L.

(v) *Average number of clusters of size s (per lattice site)* $n_s(p)$. Since sn_s is the probability that a given site is part of an s-cluster, a mean cluster size $S_p(p)$ can be defined by

$$S_p(p) = \frac{\Sigma_s s^2 n_s}{\Sigma_s s n_s}. \tag{2.2}$$

(vi) *Effective electrical conductivity* g_e. This is the electrical conductivity of a random resistor network in which a fraction p of bonds are conducting and the rest are insulating. Similarly, if a network represents the pore space of a porous medium in which a fraction p of the pores are open to flow or diffusion, an effective diffusivity D_e and a hydrodynamic permeability k can also be defined.

(vii) *Effective elastic modului G*. These are the elastic moduli of the network in which a fraction p of the bonds are elastic elements (e.g., springs), while the rest have no rigidity or stiffness (i.e., they are cut). Figure 2.2 shows

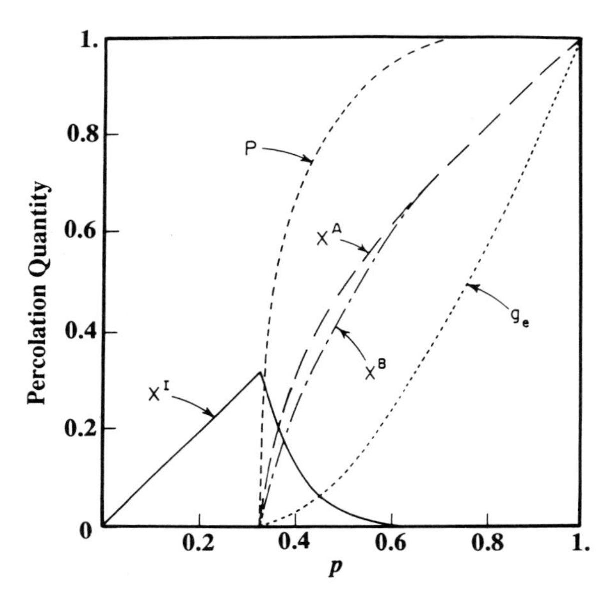

Figure 2.2 *The dependence of some of the percolation quantities on p, the fraction of occupied sites, in site percolation on a simple cubic lattice.*

the typical behavior of some of the percolation properties for a simple cubic network in site percolation, where $X^I(p)$ is the fraction of isolated occupied sites, i.e., $X^I(p) = p - X^A(p)$. The topological properties such as the accessible or backbone fractions are usually calculated by Monte Carlo simulations. For example, Stauffer *et al.* (1982) give a computer program for calculating $X^A(p)$, while Liem and Jan (1988) discuss a method for calculating $X^B(p)$. The transport properties, such as the effective conductivity or the elastic moduli, can be calculated by Monte Carlo simulations or by the analytical approximations discussed in Chapter 5.

2.4 The structure of the backbone at or near percolation threshold

Of particular interest is the structure of the backbone near or at the percolation threshold, since the backbone plays a fundamental role in any transport process in percolating systems. Following Stanley (1977), we can divide the bonds of the backbone into two groups. In one group are those that are in the *blobs*, i.e., the multiply connected part of the backbone. In the second group are the *red bonds*, those that, if cut, would split the backbone into two parts. The reason for calling such bonds red is that in a percolating electrical network they carry all of the current between two blobs, and thus they are the *hottest* bonds in the backbone.

2.5 Universal scaling laws for percolation quantities

The numerical value of every percolation quantity for any p depends on the microscopic details of the system, such as its coordination number. But near the bond or site percolation threshold p_c, most percolation quantities obey scaling laws that are largely insensitive to the network structure and its microscopic details. The quantitative statement of this insensitivity is that, near p_c we have the following scaling laws

$$P(p) \sim (p - p_c)^{\beta_p}, \tag{2.3}$$
$$X^A(p) \sim (p - p_c)^{\beta_p}, \tag{2.4}$$
$$X^B(p) \sim (p - p_c)^{\beta_B}, \tag{2.5}$$
$$\xi_p(p) \sim |p - p_c|^{-\nu_p}, \tag{2.6}$$
$$S_p(p) \sim |p - p_c|^{-\gamma_p}, \tag{2.7}$$
$$g_e(p) \sim (p - p_c)^{\mu}, \tag{2.8}$$
$$G(p) \sim (p - p_c)^{f}. \tag{2.9}$$

The topological exponents β_B, β_p, ν_p and γ_p are completely *universal*, i.e., they are independent of the microscopic details of the system, and depend only on the dimensionality of the system. Even long, but finite, range correlations do not change this universality, although they do change the value of p_c. The transport exponents μ and f are also largely universal. Later in this chapter we discuss the conditions under which the universality of μ and f may be violated.

The scaling behavior of the effective diffusivity D_e is related to that of $g_e(p)$. According to Einstein's relation, g_e and D_e are related through, $g_e \sim n_e D_e$, where n_e is the density of the electrons. Diffusion can take place both on the sample-spanning cluster as well as the finite clusters. But in most cases, only diffusion on the sample-spanning cluster contributes significantly to the long-time transport properties of the system (the case in which the finite clusters can also contribute significantly is discussed in Chapter 9). In this case, $n_e \sim X^A(p)$, and therefore

$$D_e(p) \sim (p - p_c)^{\mu - \beta_p} \tag{2.10}$$

Similarly, near p_c the permeability k of a percolating network obeys the following scaling law

$$k(p) \sim (p - p_c)^e. \tag{2.11}$$

For network models, $e = \mu$. But for percolating continua, e can be significantly different from μ.

Two other transport properties that will be used in this book are as follows. Imagine that in a percolation network a fraction p of bonds are perfect conductors (their resistance is *zero*), while the rest are normal conductors (their resistance is *finite*). Then, *below* p_c only finite clusters of perfect conductors are formed and g_e is finite. As p_c is approached from below, clusters of perfect conductors become larger and g_e increases. Finally, g_e diverges at p_c such that near p_c we have

$$g_e(p) \sim (p_c - p)^{-s}. \tag{2.12}$$

In two dimensions, $\mu = s$, but there is no known relation between s and μ in three dimensions. Similarly, if a fraction p of the bonds are totally *rigid* springs (their spring constant is *infinite*), while the rest are *soft* (their spring constant is *finite*), then the elastic moduli are finite below p_c, but diverge as p_c is approached from below as

$$G \sim (p_c - p)^{-\zeta}. \tag{2.13}$$

The relevance of s and ζ to practical applications is discussed in Chapters 11 and 12. Obviously, when the bonds are elastic springs, one can invent many types of percolation networks by simply changing the force laws that govern the deformation of the springs, and thus the numerical values of the elastic moduli would depend on the type of the force laws that are used. We show in Chapter 4 that such elastic networks are relevant to modeling of fracture of disordered solids and rock. In Chapters 11 and 12 we demonstrate the relevance of such elastic networks to modeling the viscosity and elastic moduli of polymer networks and other disordered materials.

For large clusters near p_c, $n_s(s)$ obeys the following scaling law

$$n_s \sim s^{-\tau_p} f[(p - p_c)s^{\sigma_p}], \tag{2.14}$$

where τ_p and σ_p are also universal and $f(0)$ is not singular. The geometrical exponents are *not* all independent. For example, one has, $\tau_p = 2 + \beta_p\sigma_p$ and $\nu_p d = \beta_p + 1/\sigma_p = 2\beta_p + \gamma_p$, and in fact knowledge of ν_p and another exponent is sufficient for determining most of the geometrical exponents. Unlike these exponents, the implied prefactors in all of the above scaling laws *do* depend on the type of network, which is why the *numerical* values of percolation quantities depend on the details of the system. If two phenomena are described by two different sets of critical exponents, they are said to belong to two different *universality classes*, in which case the physical laws governing the two phenomena must be fundamentally different. Thus, critical exponents can help one to distinguish between different classes of problems and the physical laws that govern them. An accurate technique for estimating these exponents is the finite-size scaling method discussed below. In Table 2.3 the values of the critical exponents in two and three dimensions are compiled. For comparison, the corresponding values for the Bethe lattices are also given.

Table 2.3 *Currently accepted values of the critical exponents and fractal dimensions for a d-dimensional system. Rational or integer values are exact results*

	$d = 2$	$d = 3$	Bethe lattices
β_p	5/26	0.41	1
β_{BB}	0.48	1.05	2
ν_p	4/3	0.88	1/2
γ_p	43/18	1.82	1
σ_p	36/91	0.45	1/2
τ_p	187/91	2.18	5/2
D_c	91/48	2.52	4
D_{BB}	1.64	1.8	2
$D_{\min}$	1.13	1.34	2
μ	1.3	2.0	3
s	1.3	0.73	0
f	see Chapter 11		
ζ	see Chapter 11		

2.6 Fractals and percolation

For any length scale $L >> \xi_p$, a percolating system is macroscopically homogeneous. But for $L << \xi_p$, the system is *not* homogeneous and the macroscopic properties of the system depend on L. In this regime, the sample-spanning cluster is on average self-similar, i.e., it looks the same at all length scales up to ξ_p, and its mass M (its total number of bonds or sites) scales with ξ_p as

$$M \sim \xi_p^{D_c}, \tag{2.15}$$

where D_c is called the fractal dimension of the cluster. However, D_c is not a totally new quantity and is given by

$$D_c = d - \frac{\beta_p}{\nu_p}. \tag{2.16}$$

For $L >> \xi_p$, $D_c = d$. Similarly, for $L << \xi_p$, the backbone is a fractal object and its fractal dimension D_{BB} is given by

$$D_{BB} = d - \frac{\beta_B}{\nu_p}. \tag{2.17}$$

Note that if $L < \xi_p$, one should replace ξ_p in (2.15) by L. Note also that ξ_p is divergent at $p = p_c$, so that then the sample-spanning cluster is a fractal object for *any* L.

For $L << \xi_p$, the number of red bonds M_{red} scales with L as, $M_{red} \sim L^{D_{red}}$, and thus D_{red} is the fractal dimension of the set of the red bonds. Coniglio (1981) proved that $D_{red} = 1/\nu_p$. Another important concept is the *minimal* or *chemical* path between two points of a percolation cluster, which is the shortest path between the two points. This concept was first discussed by Alexandrowicz (1980) and Havlin and Nossal (1984). For $L << \xi_p$ the length L_{min} of the path scales with L as

$$L_{min} \sim L^{D_{min}}, \tag{2.18}$$

and therefore D_{min} is the fractal dimension of the minimal path. Chapter 13 shows that the minimal path plays an important role in hopping conductivity of semiconductors. The current values of these fractal dimensions are also listed in Table 2.3.

Once it is established that a system is a fractal, many classical laws of physics have to be significantly modified. For example, Fick's law of diffusion with a constant diffusivity is not appropriate for describing diffusion processes in fractal systems; this is discussed in Chapters 9 and 10.

2.7 Percolation in finite systems and finite-size scaling

So far we have discussed percolation in infinitely large systems. Percolation in finite systems deserves discussion because practical applications usually involve finite systems. The effect of finite size of the system manifests itself mostly near p_c where ξ_p is large, and thus we need to study its effect in this region. Fisher (1971) investigated the effect of finite size of a thermal system on its critical properties near the critical temperature, and developed a theory for such finite systems, usually called finite-size scaling, which can be adopted for percolation processes. In a finite system, as p_c is approached, ξ_p eventually becomes comparable to the linear size of the network. Therefore, following Fisher (1971), the variations of *any* property P_L of a system of linear size L is written as

$$P_L \sim L^{-x} f(u), \tag{2.19}$$

where, $u = L^{1/\nu_p}(p - p_c) \sim (L/\xi_p)^{1/\nu_p}$, and $f(0)$ is nonsingular. If near p_c and in the limit $L \to \infty$, one has $P_\infty \sim (p - p_c)^\delta$, then one must have $x = \delta/\nu_p$. The finite size of the network also causes a shift in the percolation threshold (Levinshtein *et al.* 1976),

$$p_c - p_c(L) \sim L^{-1/\nu_p}. \tag{2.20}$$

Here p_c is the percolation threshold of the infinite system, and $p_c(L)$ its *effective* value for a finite system of linear size L. Although (2.19) and (2.20) are valid for large values of L, very large systems cannot easily be simulated in practice and therefore an equation such as (2.19) is modified to

$$P_L \sim L^{-x}[a_1 + a_2 g_1(L) + a_2 g_2(L)], \tag{2.21}$$

where g_1 and g_2 are two *correction-to-scaling* terms, particularly important for small and moderate values of L; a_1 and a_2 are constants. For *transport properties* (e.g., conductivity, diffusivity, elastic moduli), $g_1 = (\ln L)^{-1}$, and $g_2 = L^{-1}$ often provide accurate estimates of x (Sahimi and Arbabi 1991). Equation (2.21) also tells us how to estimate the critical exponents: Calculate P_L at p_c for several values of L, and fit the results to (2.21) to estimate x.

2.8 Percolation in random networks and continua

The use of regular networks for investigating various phenomena in disordered systems is popular. But percolation in continua and in topologically random networks, those in which the coordination number varies from site to site, is of great interest. This is because many practical situations deal with such systems. There are at least three ways of realizing percolating continua. For a review of continuum percolation see Balberg (1987); for the most recent references see Alon *et al.* (1991) and Drory *et al.* (1991). In the first method, one has a random distribution of inclusions, such as circles, spheres or ellipses, in an otherwise uniform system, an example of

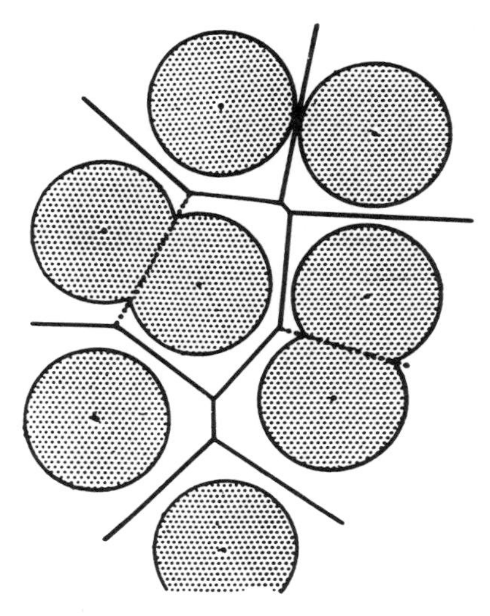

Figure 2.3 *Two-dimensional Swiss cheese model of random continua.*

which is shown in Fig. 2.3. In such systems percolation is defined either as the formation of a sample-spanning cluster of the paths between untouching inclusions, or as the formation of a sample-spanning cluster of touching or overlapping inclusions. In the second method, one divides the space into regular or random polyhedra, a fraction of which is occupied (conducting), while the rest are unoccupied; an example is shown in Fig. 2.4. In the third method, one distributes at random conducting sticks of a given aspect ratio,

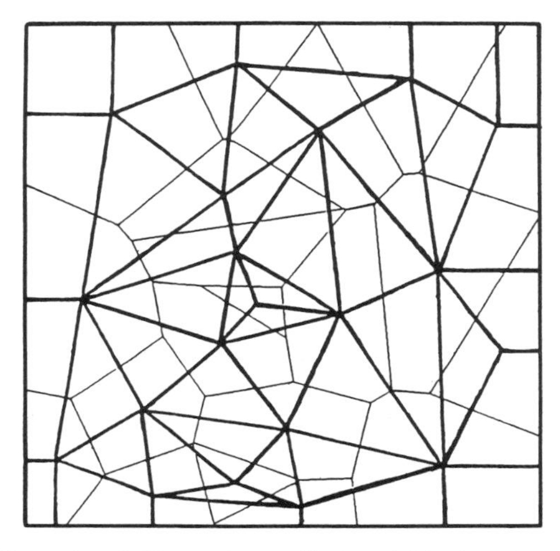

Figure 2.4 *Two-dimensional Voronoi tessellation (thin lines) and Voronoi network (thick lines).*

or plates of a given extent. The last class of disordered continua may be relevant to modeling fracture networks of rock (see Chapter 4).

More formally, one can define percolation in continua as follows. Consider a random and continuous function $h(\mathbf{r})$ at point $\mathbf{r}$, defined for the entire space, with the property that $\langle h(\mathbf{r}) \rangle = \mathbf{0}$, where $\langle . \rangle$ denotes an averaging. We now paint all regions of space for which, $h(\mathbf{r}) < R$ black, where R is a real number, while the rest are white. If R changes from $-\infty$ to $+\infty$, then the volume of the black region can vary from zero to ∞. If R is small, we can only form small black islands. But as R becomes larger, black islands can merge and, finally, at a critical R_c, form a sample-spanning black island. The function $h(\mathbf{r})$ can be thought of as a potential, such that when it reaches a certain level, the system can percolate.

One of the most important discoveries for percolating continua (Scher and Zallen 1970) is that the critical occupied volume fraction ϕ_c, which is defined as

$$\phi_c = p_{cs} f_l, \tag{2.22}$$

where f_l is the filling factor of a lattice when each of its sites is occupied by a sphere in such a way that two nearest-neighbor impermeable spheres touch one another at one point, appears to be an invariant of the system whose value is about 0.45 for $d = 2$ and 0.15–0.17 for $d = 3$. Shante and Kirkpatrick (1971) generalized this idea to permeable spheres, and showed that the average number of bonds per sites B_c at p_c (Tables 2.1 and 2.2) is related to ϕ_c by

$$\phi_c = 1 - \exp(-B_c/8), \tag{2.23}$$

and that the value of B_c for percolating continua is the limiting value of $p_{cs}Z$, when $Z \to \infty$. Accurate estimates of B_c for many systems were obtained by Haan and Zwanzig (1977). If ϕ_c is truly an invariant of continuous systems, then R_c is given by

$$\phi_c = \int_{-\infty}^{R_c} h(\mathbf{r}) dV, \tag{2.24}$$

where V is the volume of the system.

It has been established that the topological critical exponents defined above are the same for lattice and continuous systems (for a review see Balberg 1987). But *transport* (e.g., conduction) in percolating continua can be quite different from that in discrete networks. Consider, for example, the two-dimensional Swiss cheese model, shown in Fig. 2.3, in which circular inclusions (spherical inclusions in three dimensions) are punched at random in an otherwise uniform system. If transport takes place through the channels between the nonoverlapping circles, then the system can be mapped onto an equivalent percolation problem in a random network made of the *edges* of Voronoi polygons (polyhedra) (Kerstein 1983); see Fig. 2.4. If we construct the *dual* of this network, i.e., connect the centers of the neighbouring polygons (polyhedra), we obtain a Voronoi network, also shown in Fig. 2.4. The *average* coordination number of a Voronoi network is 6 in two dimensions and 15.5 in three dimensions. Kogut and Straley (1979) and Feng *et al.* (1987) used various models and techniques to show that the critical exponents μ_c, e_c, and, f_c, defined for the conductivity, elastic moduli, and permeability of continua, are quite different from μ, e, and f defined above for percolating networks. The reason is that transport in a system such as the Swiss cheese model can be dominated by the narrow necks between the nonoverlapping circles (or spheres). Since the widths of such necks can vary throughout the system, there will be a distribution of local transport properties such as the neck conductance g. If this distribution is singular near $g = 0$, then the universality of μ and other transport exponents can be violated. No such singularities appear in percolation on networks, unless the bond conductances are distributed according to such singular distributions. Feng *et al.* (1987) showed that in a three-dimensional Swiss cheese system $\mu_c \simeq \mu + 1/2$ and $e_c \simeq \mu + 5/2$. For the two-dimensional system $\mu_c = \mu$ and $e_c \simeq \mu + 3/2$.

Jerauld *et al.* (1984) showed that as long as the average coordination number of a random network and the coordination number of a regular network (e.g., the two-dimensional Voronoi and triangular networks) are about the same, many transport properties of the two systems are, for all practical purposes, identical, provided that the same bond conductance distribution is used for both networks. Therefore, in many applications, especially those that involve large-scale computer simulations, a regular network is used as it is much easier to handle. But if we use a random network to represent a disordered continuum, we need the distribution of transport properties of the bonds to closely mimic the properties of the transport paths in the continuum. On the other hand, since the topological exponents are totally universal, we can always use a random or even a regular network to study percolation in a continuum.

2.9 Conclusions

Percolation theory tells us whether a system is macroscopically connected or not. This macroscopic connectivity is of fundamental importance to many phenomena involving disordered media. Moreover, universal scaling laws near the percolation threshold tell us which aspects of a given phenomenon are important in determining its macroscopic properties, and which aspects are not relevant, and therefore we do not have to worry about them.

References

Alexandrowicz, Z., 1980, *Phys. Lett A* **80**, 284.
Alon, U., Balberg, I., and Drory, A., 1991, *Phys. Rev. Lett.* **66**, 2879.
Balberg, I., 1987, *Phil. Mag. B* **55**, 991.
Broadbent, S.R. and Hammersley, J.M., 1957, *Proc. Camb. Phil. Soc.* **53**, 629.
Coniglio, A., 1981, *Phys. Rev. Lett.* **46**, 250.
Drory, A., Balberg, I., Alon, U., and Berkowitz, B., 1991, *Phys. Rev. A* **43**, 6604.
Feng, S., Halperin, B.I., and Sen, P.N., 1987, *Phys. Rev. B* **35**, 197.
Fisher, M.E., 1971, in *Critical Phenomena: Enrico Fermi Summer School*, edited by M.S. Green, New York, Academic Press.
Fisher, M.E. and Essam, J.W., 1961, *J. Math. Phys.* **2**, 609.
Flory, P.J., 1941, *J. Am. Chem. Soc.* **63**, 3083.
Haan, S.W. and Zwanzig, R., 1977, *J. Phys. A* **10**, 1547.
Havlin, S. and Nossal, R., 1984, *J. Phys. A* **17**, L427.
Hoshen, J. and Kopelman, R., 1976, *Phys. Rev. B* **24**, 3438.
Jerauld, G.R., Scriven L.E., and Davis, H.T., 1984, *J. Phys. C* **17**, 3429.
Kerstein, A.R., 1983, *J. Phys. A* **16**, 3071.
Kogut, P.M. and Straley, J.P., 1979, *J. Phys. C* **12**, 2151.
Leath, P.L., 1976, *Phys. Rev. B* **14**, 5046.
Levinshtein, M., Shur, M.S., and Efros, E.L., 1976, *Soviet Phys.-JETP* **42**, 1120.
Liem, C. and Jan, N., 1988, *J. Phys. A* **21**, L243.
Sahimi, M. and Arbabi, S., 1991, *J. Stat. Phys.* **62**, 453.
Scher, H. and Zallen, R., 1970, *J. Chem. Phys.* **53**, 3759.
Shante, V.K.S. and Kirkpatrick, S., 1971, *Adv. Phys.* **20**, 325.
Stanley, H.E., 1977, *J. Phys. A* **11**, L211.
Stauffer, D. and Aharony, A., 1992, *Introduction to Percolation Theory*, 2nd ed., London, Taylor & Francis.
Stauffer, D., Coniglio, A., Adam, M., 1982, *Adv. Poly. Sci.* **44**, 105.
Stockmayer, W.H., 1943, *J. Chem. Phys.* **11**, 45.

3
Characterization of porous media

3.0 Introduction

In this chapter we discuss how percolation ideas have been used to study and model various morphological properties of porous media. The morphology of a porous medium consists of its geometrical properties – the shape, size, and volume of its pores – and its topological properties – the way the pores are connected to one another. There are now many sophisticated experimental techniques for characterizing the morphology of porous media. These techniques are extensively discussed by Sahimi (1993). But what we are interested in here are those experimental methods that use percolation to interpret the data. The porous media considered in this chapter can vary anywhere from porous catalysts to reservoir rocks. The structure of a porous medium depends upon its heterogeneity and the length scale at which the medium is inspected. Therefore, we restrict our attention in this chapter to porous media that are macroscopically homogeneous (although microscopically disordered), i.e., those whose properties are, for a large enough sample, *independent of their linear dimension*. In Chapters 4 and 6 we discuss macroscopically heterogeneous porous media, those that contain heterogeneities that vary in space, and whose properties may depend on the length scale of measurements. But, before we discuss porous media characterization using percolation, it may be useful to describe the processes that give rise to the present porous rocks. These are called *diagenetic processes*, and percolation ideas have also been used to model such processes.

3.1 Diagenetic processes and the formation of rocks

Rock formation starts with deposition of sediments and is followed by compaction and alteration processes that cause drastic changes in the morphology of the reservoir. Consider sandstones for example. Sandstones are assemblages of discrete grains with a wide variety of chemical compounds and mixtures. If the environment around sandstones changes, the grains start to react chemically and produce new compounds, which deposit on the surface of the sand grains. As a result, the mechanical properties of the grains, such as their resistance to fracturing, also change. The chemical and physical changes in the sand after this deposition constitute diagenetic processes. The main features of diagenetic processes are: (i) mechanical deformation of grains; (ii) solution of grain minerals; (iii) alteration of grains; and (iv) precipitation of pore-filling minerals, cements, and other materials. These features have a key influence on the volume of the content of the reservoir because they control the *porosity* ϕ of the rock, which is the volume fraction of its open pores.

Immediately after deposition diagenesis starts and continues during burial and uplift of the rock until outcrop weathering reduces it again to sediment. These changes produce an end product with specific diagenetic features, whose nature depends on the initial mineralogical composition of the system, and also on the composition of the surrounding basin-fill sediments. Given a system with a particular mineralogical composition, its diagenetic history depends on several factors which are time-dependent exposures to varying temperatures, pressures, and chemistry of the pore fluid. These factors constitute the *historical* aspects of a reservoir and affect strongly its quality. The ability of reservoir rock to produce, e.g., oil, is closely related to its diagenetic history.

Porosity of reservoir rocks is either *primary* or *secondary*. Primary porosity is due to the original pore space of the sediment. Secondary porosity is due to the fact that unstable grains or cements have undergone chemical and physical changes through reactions that form water; they have partially or entirely passed into solution. If the pore space is somehow restored, the original porosity, protected from precipitation by deposition of minerals, is converted into secondary porosity. It is believed that solution pores provide more than half of all the pore space in many sedimentary rocks. The significance of secondary porosity in sandstones was only relatively recently recognized. It is now well established that five different kinds of pores with various shapes and sizes can contain secondary porosity. We group four of these together and simply call them "pores". The fifth type, open fractures, are different from the other four; they are discussed separately in Chapter 4. The existence of secondary porosity can sometimes be recognized even with the naked eye.

These diagenetic processes lead to a morphology whose porosity is smaller than the initial porosity, and in which pores can take on essentially any shape or size. Perhaps the most important result of diagenetic processes is that pores remain interconnected even when the porosity is very low. For example, it has been found by scanning electron microscopy that some pores can be connected to up to 20 other pores. This implies that the critical porosity ϕ_c of the system for having a sample-spanning cluster of open pores (i.e., the percolation threshold of the system) is very low. Random percolation *cannot* model the formation of rock pore spaces and rock porosity. Table 2.2 shows that random percolation predicts critical porosities that are too large; more general models are needed for diagenetic processes. We consider here two such models. The interested reader can consult Schmidt and McDonald (1979) for a discussion of many aspects of diagenetic processes.

3.2 Geometrical models of diagenetic processes

How can we model diagenetic processes? Since diagenetic processes for rocks seem to be similar, and because there appear to be many similarities between the geometries of various rocks, we may hope that many fundamental elements of pore formation processes are *universal*, independent of many microscopic properties of rocks. If so, we may be able to develop a general model of pore formation and growth which can explain, at a fundamental level, many features of various rocks. A study of the literature shows that there are essentially two types of modeling approaches to this problem. The first approach, which we call *chemical* modeling, relies on the continuum equations of transport (diffusion and convection) and reaction (see Sahimi *et al.* 1990, for a review), but ignores the effect of the morphology, and in particular, connectivity, of the pore space which is the main theme of this book. The second approach is what we call *geometrical* modeling in which the details of reaction kinetics and transport processes are ignored. Instead, the diagenetic process is modeled by starting from a model of unconsolidated pore space and making several simple assumptions about the rate of change of grain and pore shapes and sizes. This approach can take into account the effect of connectivity and percolation of pores and grains. Two main models for granular media, such as sandstones, are that of Wong *et al.* (1984), usually called the *shrinking tube model*, and that of Roberts and Schwartz (1985), known as the *grain consolidation model*, which we now discuss.

In the model of Wong *et al.* (1984) we start with a network of interconnected bonds, where each bond represents a resistor whose resistance R_i is selected from a probability distribution. These resistors represent cylindrical fluid-filled tubes with random radius r_i. To mimic the consolidation process and the reduction of the porosity during the diagenetic process, a tube is selected at random and its radius is reduced by a fixed factor x.

$$r_i \rightarrow x r_i, \tag{3.1}$$

where $0 < x < 1$. This simple model cannot really simulate the effect of deposition of irregularly shaped particles in an irregularly shaped pore. But it has two attractive features: (i) it preserves, for *any* $x > 0$, the network connectivity, even when the porosity has almost vanished (thus the critical porosity is almost *zero*); and (ii) the amount of change in the pore radius r_i at any step of the simulation (time) depends on the value of r_i at that time. Both of these are also true for diagenetic processes. Wong *et al.* (1984) used this model to qualitatively explain several empirical laws which relate the electrical conductivity and permeability of a porous medium to its porosity. Such empirical laws have proven to be very accurate under a wide variety of conditions, although the reason for their accuracy is not completely understood yet. These are discussed in Chapter 5. Note that the limit $x = 0$ represents random percolation, and therefore the model is essentially a sort of generalized percolation.

In the grain consolidation model we start with a dense pack of spherical grains of random radii R (Fig. 3.1(a)). The radii of the particles are then allowed to increase in unison, as a result of which the system's porosity decreases. In the region where the spheres overlap, the grains are truncated. This can be continued to yield a series of percolating porous media with various values of porosity (Fig. 3.1(b)). For the system shown in Fig. 3.1(c), $\phi_c = 0.030 \pm 0.004$. The initial (primary) and final porosities of the system depend on how the particles are originally distributed in the system. For example, in the system shown in Fig. 3.1, the particles are initially distributed randomly in such a way that they do not overlap with each other.

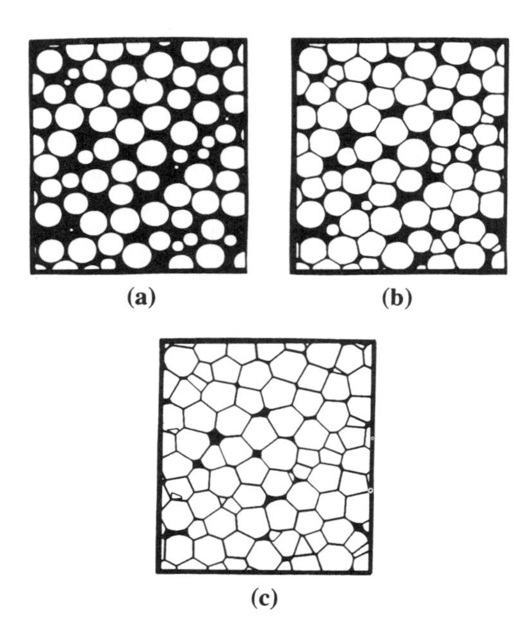

Figure 3.1 *Cross sections of three stages of the grain consolidation model of sandstone. The porosities of the systems are, (a)* $\phi = 0.364$, *(b)* $\phi = 0.2$, *and (c)* $\phi = \phi_c = 0.030$ *(after Roberts and Schwartz 1985).*

The initial porosity of this system is 0.364. On the other hand, if we start with a simple cubic lattice in which spherical grains of unit radius are placed at its nodes and follow this algorithm, then we find that $\phi_c = 0.349$, close to that for random sphere packing. But, if we start with a body-centered cubic lattice of spheres, we obtain $\phi_c \simeq 0.0055$. Therefore, the model is very flexible, and because the porosity of sandstones and similar rocks is usually less than 0.4, it provides a reasonable model of the diagenetic process. Moreover, the resulting porous media closely resemble natural sandstones. Schwartz *et al.* (1989) also considered an extension of the model in which the initial grains were not spherical but ellipsoidal, in order to model anisotropic or layered media which are abundant in nature. The diagenetic processes that give rise to many sedimentary rocks, such as sandstones, tend to favor distribution of grains that are roughly equal in size, in which case the algorithm of Roberts and Schwartz (1985) is very efficient (see Chapter 5). In Chapter 5 we discuss transport properties of porous media and compare the predictions of this model with the experimental data.

These two models have been useful in developing a unified framework for the description of many properties of granular media such as sandstones. However, such porous media possess pore or solid phases that have relatively simple characteristics. Other porous media, e.g., carbonate rocks (such as those of Iran), are more complex and their pore and solid phase geometries are not as simple as those of granular porous media. For example, most minerals in carbonate rocks are relatively soluble carbonate materials, whereas sandstone's grains originate through erosion of existing rocks with transportation of the minerals by fluid flow to the site of deposition. The grains in carbonate rocks pack more loosely than those in sandstones, and they are usually large with shapes like twigs, rods, and flakes, instead of cylinders or tubes. No percolation model of the types covered in this book has yet been proposed for such rocks.

The description of diagenetic processes and their two percolation-based models is now complete. The end product of diagenetic processes is the present porous rock. We now use percolation concepts to discuss characterization of morphological properties of porous media.

3.3 Pore space geometry: mercury porosimetry and percolation

We discuss geometrical properties of porous media and the experimental methods for measuring then. Interpretation of the data is not straightforward and requires proper modeling. We are interested only in those methods that use percolation for interpreting the data. Other methods and models are reviewed by Sahimi (1993).

One of the most important quantities for characterizing pore space geometry is the pore size distribution. But, while practically every paper and book on porous media talks about "pores" and "pore size distribution," most of them do not clearly define what is meant by either one of them. Careful examination of natural porous media reveals that what is usually referred to as pores can in fact be divided into two groups. In the first group are *pore bodies* where most of the porosity resides, while in the second group are *pore throats*, which are the channels that connect the pore bodies. We usually assign *effective* radii to pore bodies and throats which are roughly the radii of spheres that have the same volume as those of the pore body and pore throat. Porous media are often represented by a network of bonds and sites; the pore bodies are represented by network sites and the pore throats are represented by network bonds. We define the pore size distribution as the *probability density function* that gives the distribution of pore volume by an effective or characteristic pore size. Even this definition is somewhat vague. The pores are interconnected so the volume we assign to a pore can depend upon the experimental method and the model of pore space we employ to interpret the data.

Two widely used methods of measuring pore size distributions that use percolation to interpret the data are mercury porosimetry and adsorption–desorption experiments. In mercury (Hg) porosimetry, the porous medium is evacuated and immersed in Hg, which starts to penetrate the pore space if an external pressure is applied to the system. The applied pressure is then gradually increased and the volume of Hg penetrated into the porous medium is measured as a function of the pressure. Larger and larger pressures are needed to penetrate the increasingly smaller pores. The pressure is then lowered back to atmospheric pressure, as a result of which the Hg is retracted from the pores. During this process there is a characteristic shift, or *hysteresis*, between the injection and retraction curves. There is also some Hg that stays in the medium. At the end of retraction, Hg can be reinjected into the medium to obtain a second injection curve. But this is not the same as the first curve (there is hysteresis again). This technique was first developed in the 1940s, and has remained popular ever since. It is usually used for pores between 3 nm and 100 μm.

While mercury porosimetry is a relatively straightforward experiment, interpretation of the data is not simple. The data are usually interpreted using the Washburn equation

$$P_c = \frac{2\sigma_{mv}}{r}\cos(\theta + \varphi), \tag{3.2}$$

where P_c is the applied pressure, often called the capillary pressure, σ_{mv} is the interfacial tension between mercury and the vacuum, θ is the contact angle between mercury and the surface of the pores, and φ is the wall inclination angle at which the pore radius is r, with $r_t \leqslant r \leqslant r_b$, where r_t and r_b are, respectively, the pore throat and the pore body radii. Equation (3.2) results from a capillary force balance on a cylindrical tube. Dullien (1979) gives a long list of references for capillary pressure curves of various porous media.

Mercury porosimetry belongs to the general class of two-phase flows in porous media (see Chapter 7). In general, if a nonwetting fluid (in this case Hg), one for which the contact angle is larger than 90°, is to displace a perfectly wetting fluid, one for which the contact angle is nearly zero, it must overcome a capillary pressure at the pore *throat*,

$$P_c = P_n - P_w = 2\frac{\sigma_{nw}}{r_t}, \tag{3.3}$$

where P_n and P_w are the pressures in the nonwetting and wetting phases, respectively, and σ_{nw} the interfacial tension between the phases. Similarly, for the wetting phase to displace the nonwetting phase in the pore segment, the capillary pressure is related to the pore *body*

$$P_c = 2\frac{\sigma_{nw}}{r_b}. \tag{3.4}$$

Thus, in general, the shapes of capillary pressure curves can be characteristic of the pore size distribution and the wettability of the pore space. Typical curves for various wettability conditions are shown in Fig. 3.2.

Although the effect of pore space interconnectivity on mercury porosimetry, or more generally, on any two-phase flow problem in porous media, had been appreciated for a long time (Meyer 1953, Ksenzhek 1963), it was

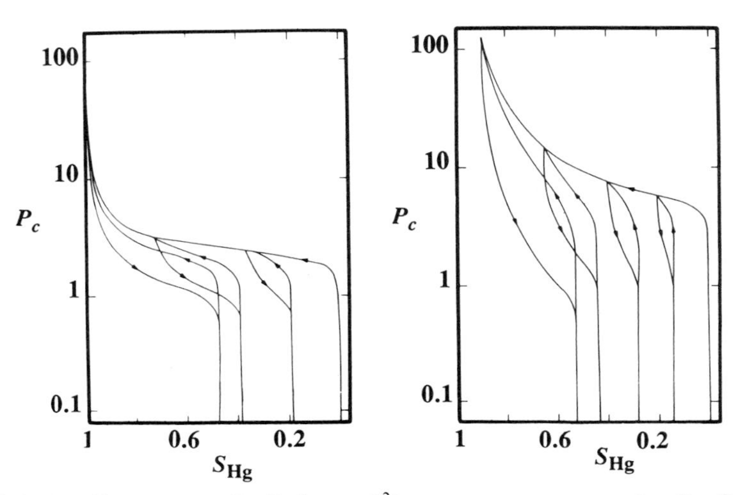

Figure 3.2 *Capillary pressure P_c (in* kg cm^{-2}*) versus mercury saturation S_{Hg} for Becher dolomite (left) with porosity $\phi = 0.174$, and Midale dolomite (right) with $\phi = 0.23$. Arrows indicate the direction of injection or withdrawal of Hg (after Larson and Morrow 1981).*

only relatively recently that the connection between this phenomenon and percolation was recognized. Chatzis and Dullien (1977) and Larson and Morrow (1981) were among the first to recognize the connection between percolation, mercury porosimetry, and capillary pressure phenomena. These authors (and many others; for a review see Sahimi 1993) recognized that a bundle of parallel capillary tubes that had been traditionally used for modeling a pore space is grossly inadequate for interpreting mercury porosimetry data. Injection and retraction processes take place in a pore space of interconnected pores, and the connectivity of the pore space greatly affects these processes. Pores that are close to the external surface of a porous medium can be reached more easily than those that are in the middle of the medium, since if a pore in the interior of the medium is to be penetrated by Hg, a connection with the external surface via the penetrated pore bodies and pore throats has to be established. If this percolation effect is not taken into account, we obtain a wrong pore size distribution.

Let us now use percolation concepts and describe a simple model for penetration of a porous medium by Hg and the properties of such a process. A percolation picture for describing any two-phase flow phenomena in porous media is appropriate if the capillary pressure across a meniscus separating the two fluids (e.g., Hg and the vacuum) is greater than any other pressure difference in the problem, e.g., due to buoyancy. The second condition is that frictional losses due to viscosity must be small compared to the capillary work. We define a capillary number Ca by

$$Ca = \frac{\eta v}{\sigma_{mv}}, \tag{3.5}$$

where v is the average fluid velocity and η is the average viscosity. Then we must have $Ca \ll 1$ in order to fulfill this criterion. The porous medium is represented as a three-dimensional network of pore bodies and pore throats. For now, we ignore the inclination angle φ, and the size of the pore bodies and consider a pore size distribution $f(r)$ for the pore throats. Because of hysteresis, during injection and retraction, the subdistributions of the pores accessible to and occupied by Hg are different from each other and different from the overall pore size distribution of the pore space. Equation (3.2) indicates that when a nonwetting fluid such as Hg invades a pore space, it first penetrates the *largest* pores (for which the required P_c is the smallest). Thus, for a given P_c, there is a minimum effective radius r_{min} such that no pore with $r < r_{min}$ is penetrated by Hg, and the fraction of pores that are *allowed* (or open) to Hg (Heiba *et al.* 1992), i.e., those pores that can be *potentially* occupied by it, is

$$X_i(r_{min}) = \int_{r_{min}}^{\infty} f(r)\,dr. \tag{3.6}$$

Therefore, X_i plays the same role as p, the fraction of open bonds, in random percolation. But only a smaller fraction of such pores can actually be reached by Hg, because some of the larger pores are connected to those with $r < r_{min}$ and cannot be reached. The eligible pores that can be reached, called *accessible* pores, will eventually form a sample-spanning cluster filled with Hg. Their accessibility is defined in the same sense as the accessibility of accessible bonds in percolation (see Chapter 2). Hence, the fraction of such pores is $X^A(X_i)$, and the distribution $f_i(r)$ of the pore radii that are occupied by Hg during injection is (Heiba *et al.* 1992)

$$f_i(r) = \begin{cases} f(r)/X^A & r \geqslant r_{min} \\ 0 & r < r_{min} \end{cases}. \tag{3.7}$$

Consider now the retraction process during which Hg is expelled from the pore space. As the pressure is lowered, Hg is first expelled from the *smallest* pores (see Fig. 3.2). The *allowed* fraction of such pores, i.e., those pores from which Hg can be *potentially* expelled, is (Heiba *et al.* 1992)

$$X_r = \int_0^{r_0} f(r)dr + \left[1 - \frac{X^A(X_{i,t})}{X_{i,t}}\right]\int_{r_0}^{\infty} f(r)dr, \tag{3.8}$$

where r_0 is the radius of the pore at a given capillary pressure P_c such that Hg is expelled from all pores with radius $r \leqslant r_0$, $X_{i,t} = X_i(r_{min,t})$, and $r_{min,t}$ is the pore radius at the end of the injection process. The first term of the right side of (3.8) is the fraction of pores from which Hg is expelled, if at the end of injection there were no pores that were not inaccessible to it. At the end of injection, a fraction $1 - X^A(X_{i,t})/X_{i,t}$ of the pores could not be reached by Hg (even though they were allowed) and, consequently, the second term of the right side of (3.8) is the fraction of pores that were not invaded by Hg at the end of injection. Hence, the size distribution of the pores from which Hg is expelled is given by

$$f_r(r) = \begin{cases} \dfrac{f(r)}{X_r}\left[1 - \dfrac{X^A(X_{i,t})}{X_{i,t}}\right] & , \quad r > r_0 \\ \dfrac{f(r)}{X_r}\left\{1 - \dfrac{X^A(X_{i,t})}{X_{i,t}}\left[1 - \dfrac{X^A(X_r)}{X_r}\right]\right\}, & \quad r_{min,t} < r < r_0 \\ \dfrac{f(r)}{X_r} & . \quad r < r_{min,t} \end{cases} \tag{3.9}$$

Let us specify clearly all the assumptions of the model: (i) the pore space is infinitely large; (ii) the entire process can be described by a random bond percolation; and (iii) entrapment of Hg in isolated clusters is ignored. The first assumption is essential if we are to use X^A for percolation processes on infinitely large lattices. But finite-size scaling (see Chapter 2) allows us to investigate systematically the effect of sample size. Strictly speaking, the second assumption is not correct. It is true the injection process is controlled by the radii of pore throats (3.3) and therefore can be considered as a bond percolation process but the same is not true of retraction. Retraction is controlled by the size of the pore bodies (3.4) and is similar to a site percolation process. Therefore, a complete modeling of mercury porosimetry as a percolation problem should involve a mixture of bond and site percolation, and size distributions for *both* pore bodies and pore throats, whereas the above formulae are derived assuming only one size distribution for the pore space. The accessibility function for a mixed bond-site percolation is more complex than discussed in Chapter 2, and has been discussed by Parlar and Yortsos (1989).

The assumption that the entire phenomenon is a *random* percolation process is not strictly correct. In practice, the pore space is invaded by the mercury from its *external surface*, so the phenomenon is an *invasion percolation* process. Chapter 6 shows that the error caused by this assumption is often very small and can be neglected. Although the third assumption is not completely correct, the resulting error is not large. We may consider a percolation problem in which trapping of clusters of one kind is allowed if they are surrounded by clusters of another kind (Sahimi and Tsotsis 1985), but computer simulations (Dias and Wilkinson 1986) have shown that, at least for three-dimensional networks, the effect of trapping is so small that it can be neglected.

What is the effect of sample size on the capillary pressure curves? The main effect is increased accessibility of pore space, which causes reduction in the sharpness of the injection curve knee. Injection curves for unconsolidated packings indicate rather strong dependence on sample thickness for systems up to about 10 particle diameters or about 30 pore throat diameters. For thicker media, the dependence is relatively weak, and if the thickness exceeds 20 particle diameters, no appreciable sample size can be detected. Finite-size scaling can be used to investigate such effects. Larson and Morrow (1981) carried out an extensive study of the effect of sample size on capillary pressure curves using a percolation model; see also Thompson *et al.* (1987).

Given this percolation picture of mercury porosimetry, how do we correlate or predict capillary pressure curves? Consider the injection process, for which the Hg saturation S_{Hg} (i.e., the volume fraction of the pores filled with Hg) is given by

$$S_{\text{Hg}} = X^A(X_i)\,\frac{\int_{r_{min}}^{\infty} f_i(r)\,V_p(r)\,dr}{\int_0^{\infty} f(r)\,V_p(r)\,dr}, \tag{3.10}$$

where $V_p(r)$ is the volume of the pore throat of radius r (recall that pore bodies were neglected). Equation (3.10) is nothing but the weighted volume fraction of pores occupied by Hg during injection. Thus, the Hg saturation

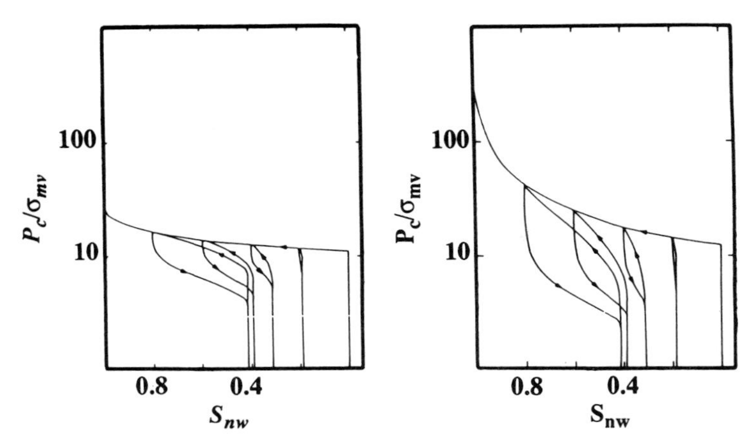

Figure 3.3 *Percolation predictions of P_c/σ_{mv} versus the nonwetting phase saturation (e.g., mercury) S_{nw} for a porous medium with a narrow (left) or broad (right) pore size distribution. Compare these with those in Fig. 3.2 (after Larson and Morrow 1981).*

during any process can be determined by a similar reasoning. Equation (3.10) tells us that a functional form for $V_p(r)$ and hence a pore shape have to be assumed. We also need to know the function X^A, which means that either the average coordination number $\bar{Z}$ of the pore space has to be known from measurements, or it must be treated as an adjustable parameter of the system in order to fit the percolation model to the data. Now, given any value of P_c and a $f(r)$, one calculates r_{min} (3.3) from which S_{Hg} is obtained. In this way the P_c vs. S_{Hg} curve is obtained. Figure 3.3 shows the predicted curves if we use a Bethe lattice of coordination number $Z = 4$ (for which X^A can be determined analytically) and $f(r) = 2r\exp(-r^2)$. They compare favorably with typical experimental data (Fig. 3.2), even though the pore space model or the pore size distribution may not seem very realistic. This is principally because the main controlling factor in mercury porosimetry – the connectivity (percolation) effect – has been explicitly taken into account. Capillary pressure curves depend on the wettability of the porous medium, and the model discussed here is appropriate for the case when one fluid is strongly wetting, while the other fluid is completely nonwetting. Heiba *et al.* (1983) also developed percolation models for the case when none of the fluids is strongly wetting (*intermediate wettability*), and the case in which a fraction of the pores are wetted by one of the fluids, while the rest are wetted by the other fluid (*mixed wettability*). We can of course use a network model and extensive computer simulation to explicitly simulate the invasion of the system by Hg, in which case considerable details can be incorporated into the model (Tsakiroglou and Payatakes 1990). But even these models make implicit use of pore connectivity and accessibility. The interested reader is referred to Sahimi (1993) where extensive references to percolation modeling of porosimetry can be found.

How can we extract the overall pore size distribution of the pore space using the percolation model? Given that $r_{min} \sim P_c^{-1}$, and that S_{Hg} and P_c are both measurable, we can assume an $f(r)$ as an initial guess and iterate (3.10) many times until it is satisfied. Normally we assume a particular form of $f(r)$ with one or two adjustable parameters. The parameters are varied until a satisfactory fit is found. But for $p < p_c$, $X^A = 0$, so we *cannot* obtain the complete pore size distribution; nor can we get information about the largest pores that are penetrated by the mercury. If we use the retraction data, we obtain information about the size distribution of the pore bodies but this too is incomplete. One way of resolving this difficulty is to do the measurements in small samples which have small percolation thresholds, as a result of which more information could become available. We need to take care to ensure that the small sample used is representative of the actual porous medium.

3.4 Pore space geometry: adsorption–desorption and percolation

Another method of determining the pore size distribution of a porous medium is using adsorption–desorption isotherm data. Liquid nitrogen is normally used in such an experiment although, in principle, gases can also be used. Let us consider first nitrogen adsorption in a single pore in which the pressure is increased, as a result of which an adsorbed film of nitrogen forms on the pore walls whose thickness increases with increasing pressure. At condensation pressure P_{co} the pore is filled with a (liquid-like) condensed phase, which results in a step increase in the adsorption isotherm. The condensation pressure is given by the Kelvin equation for a pore of radius r

$$\frac{P_{co}}{P_0} = \exp[-2\sigma_{lv} V_L/(RTr)], \tag{3.11}$$

in which P_0 is the saturation pressure, σ_{lv} the liquid–vapor surface tension, R the gas constant, T the temperature, and V_L the molar volume of the liquid. Thus, at any P_{co}/P_0 the adsorption process can be uniquely parameterized by an effective radius, from here on denoted r_a. Adsorption processes correspond to an increase in r_a; desorption processes correspond to a decrease in r_a. During adsorption, all pores are equally accessible, vapor condenses in all pores of size $r > r_a$, and liquid nitrogen fills the pores. For $r < r_a$, pores fill rapidly and continuously with nitrogen. Thus, during this adsorption process, often called *primary* adsorption, connectivity of the pores plays no role. All that matters is the effective size of the pores.

Consider now the primary desorption process. As the pressure is reduced, the desorption isotherm does not retrace that of adsorption but, similar to mercury porosimetry, forms a hysteresis loop before rejoining the adsorp-

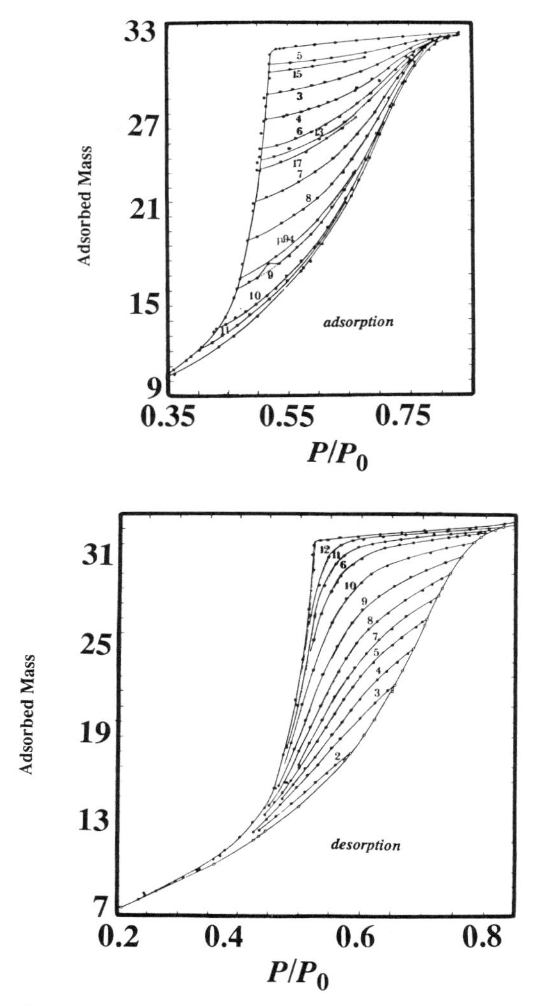

Figure 3.4 *Typical adsorption–desorption isotherms. The adsorbed mass is in mg (after Mason 1988).*

tion isotherm. Unlike the primary adsorption process, the geometry and interconnectivity of a pore do matter. A pore with an effective radius r is allowed to desorb (to contain vapor) if $r > r_a$, *and* if it has access to either the bulk vapor in primary desorption, or the isolated vapor pockets in secondary desorption, which occurs after the secondary adsorption. Typical adsorption–desorption isotherms are shown in Fig. 3.4. The primary desorption isotherm is similar to the percolation curves shown in Chapter 2.

Somewhat similar to the injection stage of mercury porosimetry, desorption is controlled by pore throats. Let $f_{pt}(r)$ be the size distribution of the pore throats. Desorption starts at the percolation threshold at which a sample-spanning cluster of pore throats containing vapor is formed. Therefore, using (3.6), the onset of primary desorption is defined in terms of a radius r_i such that

$$\int_{r_i}^{\infty} f_{pt}(r)\,dr = p_{cb}. \tag{3.12}$$

On the other hand, adsorption is controlled by the pore bodies. Let $V_{pb}(r)$ be the volume of a pore body and $f_{pb}(r)$ its size distribution. Percolation and connectivity play no role during the primary adsorption, so the saturation $S_{L,A}$ of the liquid during adsorption is given by

$$S_{L,A} = 1 - \frac{\int_{r_a}^{\infty} f_{pb}(r)\,V_{pb}(r)\,dr}{\int_0^{\infty} f_{pb}(r)\,V_{pb}(r)\,dr}. \tag{3.13}$$

Note the similarity between (3.10) and (3.13).

Primary desorption begins at the end of primary adsorption. During this process, a pore filled with liquid vaporizes if $r > r_a$, and if it is accessible to a sample-spanning vapor phase. Thus, the fraction of pore bodies or pore throats actually occupied by the vapor is given by

$$X_j = X_j^A \quad j = pb,\, pt, \tag{3.14}$$

where X_j^A is the percolation accessibility function (Chapter 2). The size distribution of the liquid-filled pores is simply given by

$$f_{L,j}(r) = \begin{cases} f_j(r)/(1 - X_j) & , \quad r < r_a \\ f_j(r)(1 - X_j/p_j)/(1 - X_j), & r > r_a \end{cases} \qquad j = pb,\, pt \tag{3.15}$$

where the quantity p_j is given by

$$p_j = \int_{r_a}^{\infty} f_j(r)\,dr, \quad j = pb,\, pt, \tag{3.16}$$

and is simply the fraction of pore bodies or pore throats that have a radius greater than r_a. The corresponding liquid saturation during desorption is

$$S_{L,D} = (1 - X_{pb})\,\frac{\int_0^{\infty} f_{L,pb}(r)\,V_{pb}(r)\,dr}{\int_0^{\infty} f_{pb}(r)\,V_{pb}(r)\,dr}. \tag{3.17}$$

Similar equations can be derived for secondary adsorption and desorption processes (see Parlar and Yortsos 1988). Similar to mercury porosimetry, (3.15) and (3.17) provide methods for correlating the data and determining $f_{pb}(r)$. We assume a functional form for $V_{pb}(r)$ and use an estimate of the average coordination number $\bar{Z}$ to determine the function X_j^A. If we ignore pore bodies and attribute everything to the pore throats, then the equations for desorption can be used for obtaining a pore size distribution for the pore space. If we assign effective sizes to both pore bodies and pore throats then, since the effective radii of all pore throats that are connected to the same pore body must be smaller than the effective radius of the pore body itself, the size distributions of the pore bodies and pore throats must obey certain restrictions (this is also true for mercury porosimetry). The connection between percolation and desorption was first noted by Wall and Brown (1981). Many other authors have discussed and investigated this connection; see Sahimi (1993) for a review.

In addition to pore size distribution, the geometry of a pore space is also characterized by its porosity and the roughness of the surfaces of its pores. It is now well-established that, over certain length scales, pore surface and pore volume of most porous media are fractal, although their fractal dimensionalities are probably not related to those of percolation. The fractality of pore surfaces and pore volumes was demonstrated in a series of beautiful papers by Katz, Thompson, and coworkers and by many other authors. These works have been reviewed extensively by Thompson *et al.* (1987) and Sahimi (1993). Based on the fractal picture, it has been suggested that

$$\phi = c\left(\frac{l_1}{l_2}\right)^{3 - D_p}, \tag{3.18}$$

where c is a constant of order unity, l_1 and l_2 are the lower and upper length scales for fractal behavior, and D_p is the fractal dimension of the pore space. The predictions of this equation agree well with the measured values for many porous media. Pfeifer *et al.* (1984) proposed that the total volume V of pores of diameters $\geqslant 2r$ obeys

$$-\frac{dV}{dr} \sim r^{2 - D_s}, \tag{3.19}$$

where D_s is the fractal dimension of the pore surface. This equation can thus be used to extract a pore size distribution for the pore space. Pore surface areas can also be measured (using, e.g., adsorption methods) and correlated with D_s; see Pfeifer *et al.* (1984).

3.5 Pore space topology: adsorption–desorption and percolation

One of the simplest concepts for characterizing the topology of a porous medium is the coordination number Z, loosely defined as the number of pore throats connected to a pore body of the medium. For regular pore structures, such as cubic arrays of spheres, it is easy to determine Z, whereas estimating Z for an irregular pore space is usually difficult and often ambiguous. We have to define an average coordination number $\bar{Z}$, and the averaging has to be taken over a large enough sample. For microscopically disordered, macroscopically homogeneous media, $\bar{Z}$ is independent of sample size.

How can we determine the average coordination number and other topological properties of a porous medium? Many methods of estimating such properties have been reviewed by Sahimi (1993). What we are interested in here is a method that uses percolation to interpret data and estimate, e.g., $\bar{Z}$. Mason (1988), Seaton (1991), and Liu *et al.* (1992) have developed a method that uses percolation concepts and adsorption–desorption data for estimating $\bar{Z}$. We follow Seaton (1991) in describing this method.

This method is based on finite-size scaling analysis discussed in Chapter 2. We write

$$X^A(p) = L^{-\beta_p/\nu_p} f[(p - p_c)L^{1/\nu_p}], \tag{3.20}$$

which can be rewritten as

$$\bar{Z}X^A(p) = L^{-\beta_p/\nu_p} f[(\bar{Z}p - B_c)L^{1/\nu_p}], \tag{3.21}$$

exploiting the fact that $B_c = \bar{Z}p_{cb}$ is almost an invariant of the system (see Chapter 2). Consider now, as an example, the desorption curve which has three typical segments (see Fig. 3.4). In the first segment we have an almost linear isotherm, because of decompression of the liquid nitrogen in the pores. In the corresponding percolation network, p, the fraction of open pores (i.e., those in which the nitrogen pressure is below the condensation pressure) increases, but X^A is still zero, because a sample-spanning cluster of open pores has not been formed yet. At the end of this interval, the network reaches p_c, a sample-spanning cluster of open pores is formed, and the metastable liquid nitrogen in the pores of the cluster vaporizes. If we further decrease the pressure, we increase the number of pores containing metastable nitrogen and the number of pores whose nitrogen has vaporized. Around the knee of the curve, almost all pores in which the nitrogen pressure is below their condensation pressure can also vaporize, and therefore $X^A \simeq p$.

Thus, this method consists of two steps: (i) $X^A(p)$ is determined from the adsorption–desorption data; and (ii) $\bar{Z}$ and L are determined by fitting (3.21) to $X^A(p)$. Now we need to assume a relation between the pore radius and length. For example, we may assume that the length and the radius of a pore are uncorrelated. The quantity $X^A(p)/p$, i.e., the ratio of the number of pores in the percolation cluster to the number of pores below their condensation pressures, can also be written as N_p/N_b, where N_b is the number of moles of nitrogen which would desorb if all the pores containing nitrogen below its condensation pressure had access to the vapor phase, and N_p is the number of moles of nitrogen which actually have desorbed at that pressure. Let N_A be the number of moles of nitrogen present in the pores at a given pressure during the adsorption experiment; let N_D be the number of moles of nitrogen present in the pores at that pressure during the desorption experiment, and let N_F be the number of moles of nitrogen which would have been present in the pores at that pressure during the desorption experiment if no nitrogen had vaporized from the pores which contain nitrogen below its condensation pressure. Then $N_p = N_F - N_D$ and $N_b = N_F - N_A$, and

$$\frac{X^A(p)}{p} = \frac{N_F - N_D}{N_F - N_A}, \tag{3.22}$$

so that $X^A(p)/p$ is written in terms of measureable quantities. The final step is to determine p, so that $X^A(p)$ can be calculated from (3.22). But this is straightforward because for a given pressure p_g we have from (3.6)

$$p = \int_r^\infty f(x)dx, \tag{3.23}$$

where r is the pore radius in which nitrogen condenses at P_g. Therefore, given the pore size distribution $f(r)$, we can determine p and $X^A(p)$. Having determined $X^A(p)$, we can use (3.21) to estimate $\bar{Z}$. Seaton (1991) used this method for estimating the average coordination number of catalyst particles; the results are in satisfactory agreement with the data.

3.6 Conclusions

Percolation provides a tool for interpreting experimental data and gaining insight into the structure of porous media. It tells us that old models of porous media, based on bundles of parallel capillary tubes are totally inadequate for interpreting mercury porosimetry and adsorption–desorption data, and they lead to serious errors and wrong conclusions. Percolation theory opens the way for a comprehensive and meaningful modeling of porous media and any phenomena that take place in them.

References

Chatzis, I. and Dullien, F.A.L., 1977, *J. Can. Pet. Tech.* **16**, 97.
Dias, M.M. and Wilkinson, D., 1986, *J. Phys. A* **19**, 3131.

Dullien, F.A.L., 1979, *Porous Media: Fluid Transport and Pore Structure*, New York, Academic Press.
Heiba, A.A., Sahimi, M., Scriven, L.E. and Davis, H.T., 1992, *SPE Reservoir Engineering* **7**, 123.
Heiba, A.A., Davis, H.T. and Scriven, L.E., 1983, *SPE Paper 12172*, San Francisco, CA.
Ksenzhek, O.S., 1963, *J. Phys. Chem. (USSR)* **37**, 691.
Larson, R.G. and Morrow, N.R., 1981, *Powder Tech.* **30**, 123
Liu, H., Zhang, L., and Seaton, N.A., 1992, *Chem. Eng. Sci.* **47**, 4393.
Mason, G., 1988, *Proc. Roy. Soc. Lond.* **A415**, 453.
Meyer, H.I., 1953, *J. Appl. Phys.* **24**, 510.
Parlar, M. and Yortsos, Y.C., 1988, *J. Colloid Interface Sci.* **124**, 162.
Parlar, M. and Yortsos, Y.C., 1989, *J. Colloid Interface Sci.* **132**, 425.
Pfeifer, P., Avnir, D., and Farin, D., 1984, *J. Stat. Phys.* **36**, 699.
Roberts, J.N. and Schwartz, L.M., 1985, *Phys. Rev. B* **31**, 5990.
Sahimi, M., 1993, *Rev. Mod. Phys.*
Sahimi, M., Gavalas, G.R. and Tsotsis, T.T., 1990, *Chem. Eng. Sci.* **45**, 1443.
Sahimi, M. and Tsotsis, T.T., 1985, *J. Catal.* **96**, 552.
Schmidt, V. and McDonald, D.A., 1979, *Society of Economic Paleontologists and Mineralogists Special Publication No. 26*, p. 209.
Schwartz, L.M., Banavar, J.R., and Halperin, B.I., 1989, *Phys. Rev. B* **40**, 9155.
Seaton, N.A., 1991, *Chem. Eng. Sci.* **46**, 1895.
Thompson, A.H., Katz, A.J., and Krohn, C.E., 1987, *Adv. Phys.* **36**, 652; *Phys. Rev. Lett.* **58**, 29.
Tsakiroglou, C.D. and Payatakes, A.C., 1990, *J. Colloid Interface Sci.* **137**, 315.
Wall, G.C. and Brown, R.J.C., 1981, *J. Colloid Interface Sci.* **82**, 141.
Wong, P.-Z., Koplik, J., and Tomanic, J.P., 1984, *Phys. Rev. B* **30**, 6606.

4
Earthquakes, and fracture and fault patterns in heterogeneous rock

4.0 Introduction

The discussion of characterization in Chapter 3 is limited to porous media that are microscopically disordered, but macroscopically homogeneous, i.e., those whose properties, on a large enough scale are independent of their linear size or extent. Conversely, macroscopically heterogeneous porous media exhibit large-scale spatial variations of their properties. Many practical applications deal with such media, so they too are considerably important. A complete description of all properties of such systems is well beyond the scope of this book. The interested reader is referred to Haldorsen *et al.* (1988) for a fuller exposition to this important subject. Here, we restrict our attention to morphological properties of the largest-scale heterogeneities in rock that interfere with fluid flow – fractures and faults – and discuss how percolation may be relevant to their description. The effect of such heterogeneities is so severe that many of the smaller scale heterogeneities, such as those at the pore or laboratory scales, may seem simple when compared to them. We also discuss the possible relevance of percolation to the occurence of earthquakes and the spatial distribution of their hypocenters. We provide evidence that the spatial distribution of earthquake hypocenters may closely be related to the structure of fracture and fault patterns of rock and to percolation.

The presence of fractures, natural or man-made, is crucial to the economics of oil production from underground reservoirs, extraction of heat and vapor from geothermal reservoirs for use in power plants, and development of groundwater resources. In all cases, fractures provide high permeability paths for fluid flow in reservoirs that are otherwise of very low permeabilities and porosities, and would not be able to produce at high rates.

With the increasing importance of groundwater pollution, there is an even greater need to better understand the structure of fracture networks. Fault patterns, on the other hand, which are closely related to fracture networks, play a fundamental role in generating earthquakes. Despite their obvious significance, characterization of fracture and fault patterns is not as well developed as characterization of porous media without fracture. But this is changing very fast. Modern ideas, such as fractals and percolation concepts, are beginning to find their proper place in this field.

Chapter 3 briefly discusses diagenetic processes for sandstones and carbonate rocks. Such porous media do not usually contain large fractures. The sedimentologic, tectonic, and diagenetic histories of fractured rock are complex and very different from conventional porous media. This gives rise to significant differences between the two. Fracture porosities are generally in the range 1–6%, whereas the pore porosity is usually larger than 10%. Because rational development of fractured rocks and maximum recovery of their contents are closely related to an accurate representation of their internal structure, development of an accurate three-dimensional map of fracture distribution is essential for any meaningful study of such complex reservoirs. This chapter discusses characterization of topological properties of fracture and fault patterns. Chapter 5 discusses fluid flow through a fracture network.

4.1 Morphological properties of fracture and fault networks

Let us now summarize the most important parameters for characterizing the morphology of a fracture network, then discuss past studies that provide strong evidence for the connection between fracture networks and fault patterns of heterogeneous rocks and percolation.

Fracture aperture is the crucial parameter which determines its permeability; the volumetric flow rate q through a fracture is proportional to its aperture cubed (for a pore, q is proportional to the fourth power of its effective radius). For rough fractures, the dependence is more sensitive, depending on powers of the aperture as high as six. It has been found that the frequency of inverse aperture, when plotted against the inverse aperture, follows a power law, a strong indication of self-similar and fractal behavior.

Fracture density and spatial geometry are both important parameters in reservoir modeling. The areal fracture density is defined as the sum of fracture trace lengths per unit area. For an isotropic network, this is the same as fracture area per unit volume.

Fracture connectivity of a network, similar to the coordination number of a pore space, has an important effect on its properties. It can be quantified by the ratio of three types of fracture termination: (i) a *blind termination* in which a fracture ends in the rock matrix; (ii) an *abutting termination* in

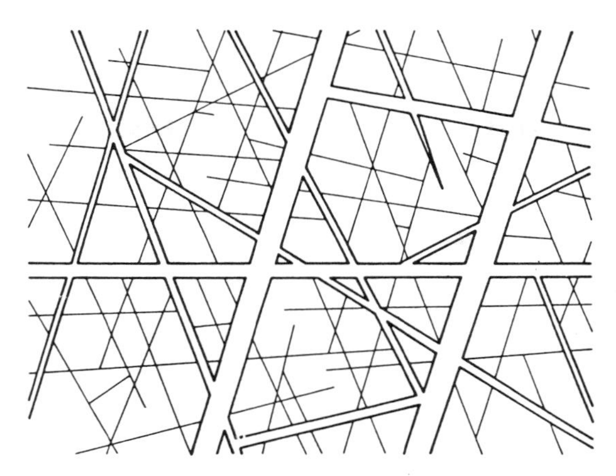

Figure 4.1 *Three types of fractures in heterogeneous rock. The fracture porosity is exaggerated.*

which a fracture ends when it intersects another fracture; and (iii) a crossing in which two fractures intersect but neither terminates (see Fig. 4.1). Among these properties, fracture connectivity and density are the most relevant to our discussions in this chapter. Fracture aperture is relevant to fluid flow in Chapter 5.

4.2 Fractal properties of fracture and fault networks and their relation with percolation

We now discuss past experimental and theoretical studies of fracture and fault patterns that indicate a strong connection between them and percolation. Many authors have studied fracture networks of heterogeneous rock masses around the world and have found them to have fractal pattern. One way of estimating the fractal dimension D_f of the pattern is to count the number N_ℓ of fractures of length ℓ, which for a fractal network follows a power law

$$N_\ell \sim \ell^{-D_f}. \tag{4.1}$$

This is called the *box counting method* of estimating a fractal dimension. It has been generally found that at *large* length scales (of the order of one kilometer or more) thin sections of fractured rock (an essentially two-dimensional system) are characterized by $D_f \simeq 1.9$, whereas at *small* scales (of the order of several meters) they are characterized by $D_f \simeq 1.6$–1.7.

One well-known example of such fractal fracture networks is the Yucca Mountain formations in Nevada, whose structure was investigated by Barton and Hsieh (1989) for the US Department of Energy as a potential

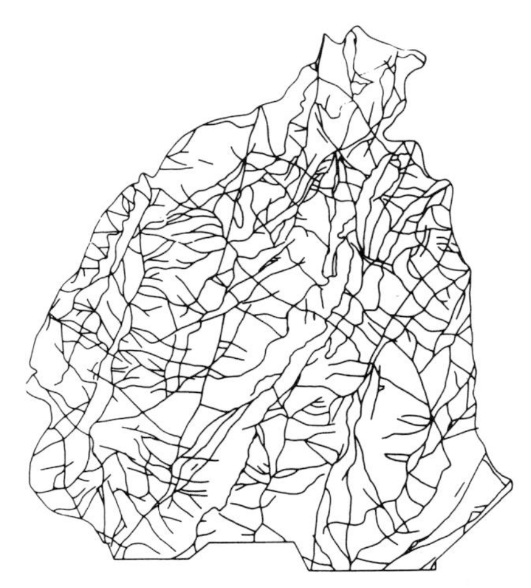

Figure 4.2 *Fracture pattern of the surface of the Geysers geothermal field (after Sahimi et al. 1993).*

underground repository for high-level radioactive wastes. Using the box counting method, they found that the fracture surface of the rock has a fractal structure, and that for fractures ranging from 20 cm to 20 m the fractal dimension D_f is about 1.6–1.7. Results similar to those for the Yucca Mountain formations have also been found for the Monterey formation in California, an important source of heavy oil, and other rocks. For some Japaness formations Hirata (1989) found that for large fracture surfaces, $D_f \simeq 1.9$, whereas for smaller scales, $D_f \simeq 1.7$.

Another important example is the Geysers geothermal field in Northeast California. It covers an area of more than 1.4×10^8 m^2 and represents one of the most important geothermal fields in the world. Sahimi *et al.* (1993) studied the fracture pattern of this field. Their results for thin (two-dimensional) sections of the field are shown in Fig. 4.2, from which $D_f \simeq 1.9$.

The above studies were restricted to fracture surfaces and thin sections. Three-dimensional fracture networks can also be studied by a few methods. In the case of the Geysers field we can obtain a reasonably accurate map of the fracture network through drilling, since large fractures, when hit by the drilling equipment, produce a sudden and measurable increase in the steam pressure on the ground. Analysing such data, it was found that as larger and larger samples of the field are examined, the fractal dimension of the fracture network increases and tends toward 2.5 for large systems.

Another way of obtaining information about three-dimensional structure of fracture and fault patterns is through seismic waves. Known in seismic literature as coda, such data are the late acoustic or seismic signals that we measure in a remote station, several seconds after the first seismic signal that has emanated from an earthquake is received by the station. Since earthquake hypocenters are located on the fault or fracture network of rock, the coda reflects the scattering of seismic waves by large-scale three-dimensional fracture or fault structures of rock. The coda data can be analysed in two different ways. In the first method, the backscattering coefficient g_{bs} is estimated from

$$g_{bs}(\omega) = \frac{v_s t_f^2 P_{co}(\omega)}{2P_s(\omega)} \exp[\omega t_f / Q(\omega)], \tag{4.2}$$

where v_s is the average wave velocity, t_f is a fixed time, and ω is the frequency. There are three measurable quantities in (4.2): $Q(\omega)$, the *quality factor* of the medium; P_{co}, the measured coda power; and $P_s(\omega)$, the source power spectrum. $Q(\omega)$ represents the intrinsic and scattering attenuation effects and is obtained experimentally from the slope of coda envelopes; Because $P_s(\omega)$ varies greatly for each earthquake, (4.2) is averaged over many earthquakes. If g_{bs} varies with ω with a power law

$$g_{bs}(\omega) \sim \omega^H, \tag{4.3}$$

then the fractal dimension is given by

$$D_f = 3 - H. \tag{4.4}$$

Alternatively, we can calculate the one-dimensional power spectrum of the seismic waves from

$$W(k_\beta) = \frac{4g_{bs}(\omega)}{Z_\beta^2} k_\beta^2, \tag{4.5}$$

where $k_\beta = \omega/v_s$, and Z_β is given by

$$Z_\beta = \left\langle \frac{\delta Z}{Z_0} \right\rangle^{1/2}, \tag{4.6}$$

and Z_0 and δZ are, respectively, the average value and the perturbation of the medium impedance. The quantity Z_β is roughly constant for many different rocks. If $W(k_\beta)$ scales with k_β as

$$W(k_\beta) \sim k_\beta^\delta, \tag{4.7}$$

then the fractal dimension is given by, $D_f = 1 + \delta$. Wu and Aki (1985) analyzed several sets of coda data and found evidence of fractal behaviour for three-dimensional large scale heterogeneities in the lithosphere with, $D_f \simeq 2.46$–2.52. Finally, Chelidze and Gueguen (1990) studied the three-dimensional fracture patterns of Stockbridge dolomite marble (in Connecticut) and found $D_f \simeq 2.5$.

So far we have discussed experimental data for fracture patterns of heterogeneous rock. These data indicate that at *large length scales* the fracture pattern of rock is fractal with a fractal dimension indistinguishable from those of percolation clusters, namely, $D_f \simeq 1.9$ and 2.5 in two and three dimensions, respectively. Later we explain why the fractal dimension of the fracture network of rock at small scales is not the same as the fractal dimension of percolation clusters. But now we discuss some theoretical evidence in support of the connection between fracture patterns of rock at large length scales and percolation.

Computer simulation studies of Long *et al.* (1991) provided strong evidence for the connection between fracture and fault patterns and percolation. Suppose that we have some limited amount of information about a fractured rock, e.g., we know the amount of fluid that it can produce at certain locations, and we would like to develop a fracture network model for the rock that honors (reproduces) our limited amount of data. To do this, Long *et al.* (1991) developed a method called *simulated annealing*. Such methods were originally developed by Kirkpatrick *et al.* (1983) and others for statistical mechanical and thermal systems, where one attempts to find a configuration of the system whose energy is minimum. Starting with a particular configuration of a fracture network as the initial guess for the structure of the rock, we change systematically the connectivity of the fractures in the network and calculate the quantities for which some data are available. For example, we can start with a network in which some bonds represent the fracture through which fluid flow takes place, and the rest of the bonds are closed to flow. If a particular configuration of the network reproduces the data to within a certain tolerance, then we accept that configuration as the optimum structure of the fracture network of the rock. Otherwise, we change that configuration once again and the quantities of interest are recalculated and compared with the data, and so on. Long *et al.* (1991) showed that, even if we start with a fully connected fracture network with a regular structure, the optimum configuration is always a network which, although it contains a large number of fractures, is barely connected, i.e., removal of a few fractures divides it into several large and dense but disconnected blobs of fractures, precisely the structure of the largest percolation cluster at the percolation threshold discussed in Chapter 2.

Finally, if, instead of the entire fracture or fault pattern, we analyze the surface of a *single major* fault, such as the San Andreas fault in California, we find that the fault is not a straight line but a fractal with, $D_f \simeq 1.1$–1.2 (Okubo and Aki 1987). Therefore, the accumulated experimental and theoretical studies leave little doubt that rock fractures and fault patterns are fractal, that their fractal dimension depends on the length scale of observation, and that at the *largest scales* the fracture or fault patterns may have the structure of a percolation cluster. But, can we develop a unified model which predicts all of these data? We now show that a percolation-based model may be able to do this.

4.3 Percolation models of fracture processes

We represent the rock (or its surface) by a network of interconnected bonds. In classical mechanics a solid phase is represented by a *spring*. Likewise, in our network the bonds are springs and represent the solid matrix of the rock. To include the porosity of the rock in the model, we can remove a fraction of the springs either at random or in a correlated fashion. The fraction of absent bonds is a measure of the prefracture porosity. Every site of the network is characterized by the displacement vector $\mathbf{u}_i = (u_{ix}, u_{iy}, u_{iz})$, and nearest-neighbor sites are connected by the springs. We consider here the case of brittle fracture for which a linear approximation is valid up to a threshold defined below. This means that the springs follow the laws of linear elasticity if they are not stretched more than a critical length, or if the stress that they suffer is not greater than a critical value, but break *irreversibly* if such critical values are surpassed.

We now have to specify the force laws that the springs follow. A convenient way of representing such force laws is through the elastic energy of the system. Chapter 2 shows we can use a wide variety of force laws that the springs obey. Here we consider a network whose elastic energy E is given by

$$E = \frac{\alpha}{2} \sum_{\langle ij \rangle} [(\mathbf{u}_i - \mathbf{u}_j) \cdot \mathbf{R}_{ij}]^2 e_{ij} + \frac{\beta}{2} \sum_{\langle jik \rangle} (\delta\boldsymbol{\theta}_{jik})^2 e_{ij} e_{ik}, \tag{4.8}$$

where α and β are the stretching and angle-changing force constants, respectively, $\mathbf{R}_{ij}$ is a unit vector from site i to site j, and e_{ij} is the elastic constant of the spring between i and j. Here $\langle jik \rangle$ indicates that the sum is over all triplets in which the springs j–i and i–k form an angle whose vertex is at i. The first term in (4.8) represents the contribution of the stretching forces, while the second term is due to the forces that change the angle between two springs connected to the same node. If we assume that even the angle between the springs that make a 180° angle with one another can change during deformation of the network, then (Kantor and Webman 1984)

$$\delta\boldsymbol{\theta}_{jik} = \begin{cases} (\mathbf{u}_{ij} \times \mathbf{R}_{ij} - \mathbf{u}_{ik} \times \mathbf{R}_{ik}) \cdot (\mathbf{R}_{ij} \times \mathbf{R}_{ik}) / |\mathbf{R}_{ij} \times \mathbf{R}_{ik}| & \mathbf{R}_{ij} \text{ not parallel to } \mathbf{R}_{ik} \\ |(\mathbf{u}_{ij} + \mathbf{u}_{ik}) \times \mathbf{R}_{ij}| & \mathbf{R}_{ij} \text{ parallel to } \mathbf{R}_{ik} \end{cases} \tag{4.9}$$

where, $\mathbf{u}_{ij} = \mathbf{u}_i - \mathbf{u}_j$. For *all* two-dimensional networks, (4.9) simplifies to

$$\delta\boldsymbol{\theta}_{jik} = (\mathbf{u}_i - \mathbf{u}_j) \times \mathbf{R}_{ij} - (\mathbf{u}_i - \mathbf{u}_k) \times \mathbf{R}_{ik}. \tag{4.10}$$

We can use many different forms of the elastic energy; (4.8) is only one of them. We may use a network of *beams* instead of springs (de Arcangelis *et al.* 1989); then, in addition to the stretching and angle-changing forces, the *torsional* or *bending* forces also contribute to E (since beams can actually bend). Although such a model is presumably appropriate for the case in which the system is under *compressive* forces, (4.8) is a realistic model for representing many disordered media. Moreover, fractal properties of a fracture network are *not* sensitive to whether torsional forces are present or not. Chapter 11 uses elastic percolation networks to explain and predict the experimental data on elastic properties of gel polymers.

Sahimi and Goddard (1986) first introduced elastic percolation models for simulating brittle fracture. They discussed three general classes of disorder. (1) Deletion of a fraction of the bonds, discussed above. (2) Random or correlated distribution of elastic constant e of the bonds. The idea is that in real rocks (or solid materials) the shapes and sizes of the channels through which stress transport takes place vary greatly, resulting in different elastic constants for different springs. (3) Random or correlated distribution of the critical thresholds. For example, in *shear* or *tension* the breakage of a spring can be characterized by a critical length l_c, such that the spring breaks if it is stretched beyond l_c. In *compression*, each bond (representing a beam) can be characterized by a critical torsional force, such that the beam breaks if it suffers a force or stress more than the critical value. The idea is that because of rock heterogeneities, different parts of a system can offer different resistances to fracture. For our discussion in this chapter, we combine the first and the third type of disorder. The interested reader can consult Herrmann and Roux (1990) or Sahimi (1992) for a more complete review of this important research field. Hence, we introduce a threshold value l_c for the length of each spring, which is selected according to a probability density function. *Any* distribution can of course be used, but we use the following distribution

$$P(l_c) = (1 - \gamma)l_c^{-\gamma}, \tag{4.11}$$

where, $0 \leqslant \gamma < 1$. This distribution has the advantage that by varying γ we can simulate both a narrow distribution ($\gamma \simeq 0$) and a broad one ($\gamma \simeq 1$).

We now start our fracture simulation by applying an external stress or strain to the system in a given type of experiment (shear, tension, or compression). We then calculate the nodal displacements $\mathbf{u}_i$ by minimizing the elastic energy E with respect to $\mathbf{u}_i$, i.e., by writing down $\partial E/\partial \mathbf{u}_i = 0$ for every node i. This results in a set of dN simultaneous linear equations for the nodal displacements for a d-dimensional network of N nodes. This set can be solved by either a direct method, such as Gaussian elimination, or by an iterative method, such as the conjugate gradient method, particularly suitable for use in vector computers such as Cray Y-MP. We then examine the bonds to see whether any of them has exceeded its critical threshold. Fracture is initiated by breaking the spring with the *largest* deviation from its threshold, i.e., the *weakest* spring. We then recalculate the nodal displacements for the new configuration of the network, select the next spring to be broken, and so on. If no spring meets the failure criterion, we gradually increase the external stress or strain. The computations are very intensive, because we have to compute the solution of a very large set of linear equations a very large number of times (of the order of thousands). Simulation stops when a sample-spanning fracture is formed and the system fails macroscopically. Although various properties of such fracture models have been investigated by many authors, here we are only interested in the structure of the resulting fracture network. The interested reader can consult Herrmann and Roux (1990) or Sahimi (1992) for more details on many other properties of such fracture models.

Let us now discuss the predictions of this model. The shape of the fracture network depends on the distribution of the heterogeneities. If the distributions of the thresholds and e_{ij} are narrow, then the system is essentially homogeneous. In such a system, once a microcrack is nucleated the stress or force at its tip is greatly enhanced, which means that the next microcrack will almost surely develop at the tip of the existing crack. The macroscopic fracture network formed in this process is not very complex: we have a single fracture spanning the system, plus some microcracks distributed in the system. Thus, the failure of such a system is very abrupt. That is, although most springs remain intact, a sample-spanning fracture is formed very quickly; this breaks the system into two pieces. The single fracture formed in this way is similar to the single major fault in natural rock mentioned above. On the other hand, if the system is highly heterogeneous with many weak and strong springs distributed throughout the system, microcracking can happen in many places. In this case, the development of the macroscopic fracture network is more gradual, and the fracture network has a complex structure. Thus, from a practical point of view, we should try to make our solid materials as heterogeneous as possible in order to make them more resistive to fracture.

If the distribution of the thresholds is narrow, then the sample-spanning fracture that is formed in a two-dimensional network is fractal with $D_f \simeq 1.1$–1.2, whereas if we analyzes *all* fractures (the sample-spanning fracture plus the isolated cracks), we find that $D_f \simeq 1.6$–1.7. These are in agreement with the data mentioned above. But the third fractal dimension, $D_f \simeq 1.9$, pertains to large scales and cannot be obtained with such a model and simulation. The reason is that at large scales rock is highly heterogeneous and its properties vary spatially. There are also correlations between the properties of various regions (Hewett and Behrens 1990). Such large-scale heterogeneities are absent in the network used in the simulations. To include such large-scale heterogeneities in the model, we divide the network into blocks, each of which represents a region of a heterogeneous rock. Within each block, the properties are constant (rock is homogeneous on the scale of the block size) but the effective properties vary between the blocks. If we now repeat the same type of simulations but with block heterogeneities, we find that (Sahimi *et al.* 1993), $D_f \simeq 1.85$, in good agreement with

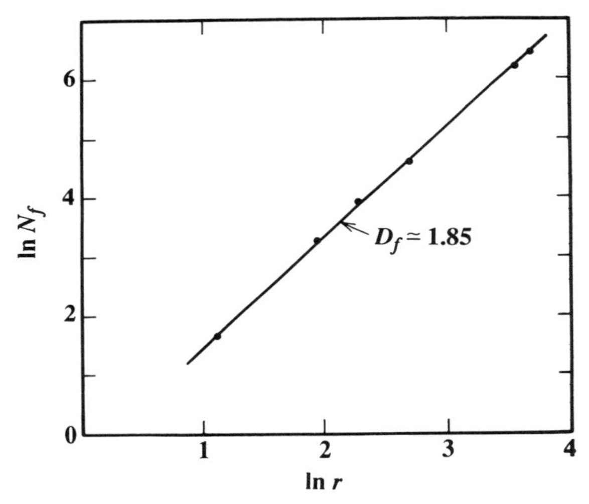

Figure 4.3 *Log–log plot of the number of fractures N_f versus the radius r obtained by the fracture model (after Sahimi et al. 1993).*

two-dimensional percolation (Fig. 4.3). Thus, it appears that fracture networks of macroscopically heterogeneous rocks are similar to percolation clusters.

Fracturing is dependent upon the stress field in the system, whereas percolation is completely random and independent of the stress field. So why are large-scale fracture networks similar to percolation clusters? Perhaps this is the explanation. Once a microcrack is initiated in a given region of the rock, it can only grow as long as it has not hit another region which is much stronger than the present region. But because the rock is heterogeneous, the growing fracture does eventually hit such a strong region, in which case its growth stops, and another crack in another region of the rock is nucleated. The growth of the new fracture also continues only as long as it has not encountered another strong region of the rock, and so on. Thus, viewed from a large scale, the nucleation and growth of fractures in highly heterogeneous rocks are more sensitive to their heterogeneities than to the stress field, and take place more or less at random, i.e., like a percolation process, *independent of the stress field.* From this perspective, fracture processes in macroscopically homogeneous and heterogeneous media are quite different, because in a homogeneous medium fractures grow almost always at their tip. It has been suggested (Sahimi and Arbabi 1992) that fracture processes can be divided into two distinct groups. One of them corresponds to fracture in a macroscopically homogeneous medium, e.g., disordered materials in which nucleation and propagation of fractures depend on the stress field in the entire system, with $D_f \simeq 1.7$ in two dimensions. The other corresponds to fracture in a heterogeneous medium, e.g., reservoir rock, which is essentially equivalent to a percolation process.

Why are such large-scale fracture networks similar to percolation clusters *at* the percolation threshold p_c but not similar to percolation clusters *above* p_c? During the initial stages of fracture propagation, large stress concentrations are generated in the system. Large stresses help the fracture growth until the fracture network becomes sample spanning; then they are released and the stress concentration becomes totally localized. This means that fracture growth cannot continue on a large scale. The significance of this is that, while fracture patterns in disordered materials are not similar to percolation clusters, large-scale fracture networks of rock can be constructed in terms of percolation clusters whose properties are well understood. This makes it much easier to model fluid flow and heat transfer in such rocks.

4.4 Generation of percolation models of fracture networks

If we accept that fracture networks in heterogeneous rocks are percolation clusters, how do we model them? Consider first a two-dimensional system. We may represent the fractures by line segments of a given length l (see, for example, Englman *et al.* 1983). As such, results of continuum percolation discussed in Chapter 2 can be used for modeling fracture networks. Alternatively, we may assign the fracture lengths from a given distribution. One of the simplest models is the Poisson model in which a $L \times L$ plane is used, and the x- and y-coordinates for centers of the fractures are selected at random from a uniform distribution in $(0, L)$. We then select the orientations of the fractures from a distribution. If the fractures cross the boundaries of the system, they are truncated. Figure 4.4 shows a typical fracture network generated by this method. The next question is how to

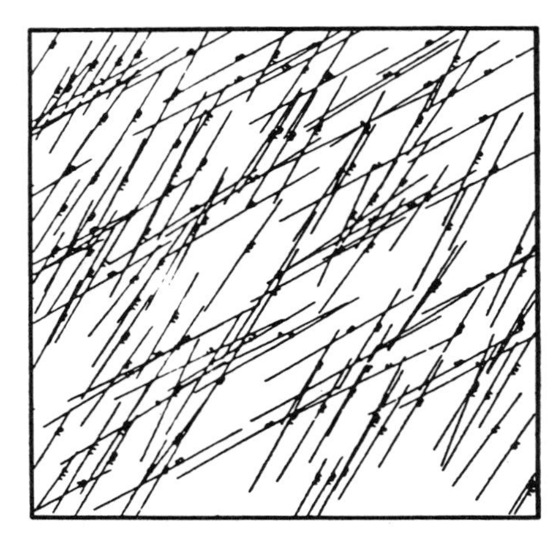

Figure 4.4 *A typical two-dimensional fracture network generated by the Poisson model.*

relate the parameters of the fracture network to those of percolation networks such as p, the fraction of conducting bonds, and $\bar{Z}$, the average coordination number of percolation networks. We may use the average number of intersections per fracture Z_f as the measure of the connectivity of the fracture network. Suppose now that the average fracture length is $\bar{l}$, the orientation distribution is $h(\theta)$, and the density of fracture centers is ρ_{cen}. Then, it is not difficult to show that

$$Z_f = \rho_{cen}\bar{l}^2 G(\theta), \tag{4.12}$$

where $G(\theta)$ is defined by

$$G(\theta) = \int_0^\pi \int_0^\pi \sin|\theta - \theta_0|\, h(\theta)h(\theta_0)\, d\theta\, d\theta_0. \tag{4.13}$$

For example, when the fracture orientation is uniformly distributed, $h(\theta) = 1/\pi$, $G(\theta) = 2/\pi$, and $Z_f = 2\rho_{cen}\bar{l}^2/\pi$. Now, for every Z_f there is a corresponding $p_f(Z_f)$, which is the analog of p for percolation, the average fraction of a fracture which is available for flow (remember the three types of fracture terminations discussed above). Consider now a fracture of constant length l with $n(l)$ intersections, and suppose that the fractures are placed in the system at random, which means that $n(l)$ is a Poisson process. Then, the average fraction of a fracture available for flow, i.e., the fraction between the two end sites on l_f is $[n(l) - 1]/[n(l) + 1]$. Therefore, if P_n is the probability that $n(l) = n$, then $P_n = Z_f^n e^{-Z_f}/n!$, and (Hestir and Long 1990)

$$p_f(Z_f) = \sum_{n=2}^{\infty} \frac{n-1}{n+2} P_n = \left(1 + \frac{2}{Z_f}\right)(1 + e^{-Z_f}) - \frac{4}{Z_f}. \tag{4.14}$$

Given this equation, all standard results of percolation discussed in Chapter 2 can now be used for fracture networks, provided that we replace p everywhere with p_f. Thus, an exact one-to-one correspondence between fracture and percolation networks can be made. For other types of fracture networks we can derive a relation between p_f and other parameters of the system; see Hestir and Long (1990). Generation of three-dimensional fracture networks can be done in a similar way, assuming a shape for the fractures. The interested reader can consult Sahimi (1993) for more discussions and references.

4.5 Earthquakes, fracture and fault patterns, and percolation

The distribution of energy released during earthquakes has been found to obey the famous Gutenberg–Richter law (Gutenberg and Richter 1956).

This law is based on the empirical observation that the number N of earthquakes of size greater than m is given by

$$\log_{10}N = a - bm, \tag{4.15}$$

where values of a and b depend on the location, but generally b is in the interval $0.8 < b < 1.5$. The energy E released during the earthquake is believed to be related to m by

$$\log_{10}E = c - em, \tag{4.16}$$

so that the Gutenberg–Richter law is essentially a power law connecting the frequency distribution function with the energy release E (or other physical quantities). If we combine (4.15) and (4.16), we obtain

$$\log_{10}N = a_1 - b_1\log_{10}E, \tag{4.17}$$

a power law describing the fact that the larger the earthquake the fewer the occurrences. Equation (4.17) is reminiscent of (2.14) for $n_s(p)$, the cluster number; and s, the cluster size in percolation theory.

Otsuka (1979) appears to be the first to attempt relating percolation properties with the empirical dependence of N on E. He argued that the surface matter of the earth's crust is always being somewhat distorted, and that earthquakes occur when this distortion exceeds a certain threshold value. Moreover, he argued that each unit volume of rock contains an energy distortion (preceding an earthquake) of $e_d \sim 2 \times 10^4\,\mathrm{erg\,cm^{-3}}$, and thus he proposed that numerically $E \simeq V$, where V is the volume of the crust about to produce an earthquake. In this way, he drew a closer analogy between occurrence of earthquakes and percolation, since (4.17) can be rewritten in terms of the number and "volume" associated with the "clusters" of the severely distorted earth's surface matter.

In Otsuka's model, the role of the probability p of percolation is played by the probability that a particular piece of the crust is severely distorted. Thus, this parameter must depend on the specific sample region and the geological conditions during the period of observation. In particular, the power law of earthquake "clusters" will *not* be observed unless $p = p_c$. By comparing the earthquake frequency data from various regions of the world, Otsuka demonstrated that this is indeed consistent with the data, i.e., the power law (4.17) is an oversimplification. His figures do seem to indicate that certain regions, e.g., northern Japan, correspond to $p < p_c$, while others, e.g., South America, perhaps to $p > p_c$. Of course, every earthquake is restricted to a *finite* region, and thus no earthquake "cluster" is treated as infinite, even if $p > p_c$. However, Otsuka's quantitative results are not directly comparable with percolation quantities discussed in Chapter 2 as he did not use proper normalization, did not use suitable quantities to plot, and used

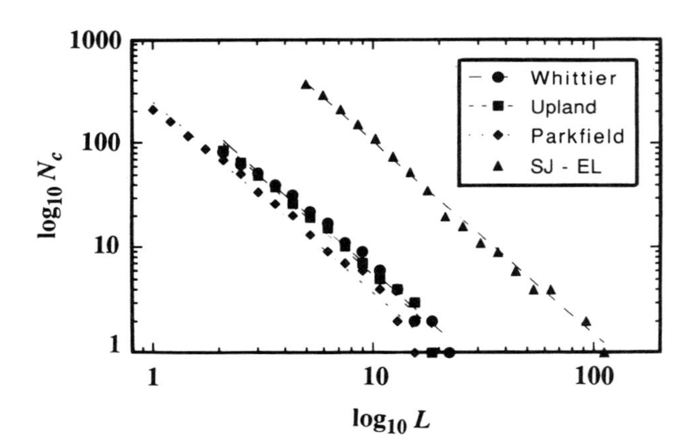

Figure 4.5 *Variations of the minimum number of occupied cubes N_c of edge L (in kilometers) to completely cover the data set at various length scales L for all four sets of earthquake data. The slopes of all data sets are about 1.8 (after Sahimi et al. 1993).*

a Bethe lattice of coordination number $Z = 3$. Chelidze (1981), Trifu and Radulian (1989), and Stark and Stark (1991) also attempted to make a connection between the earthquake statistics and percolation, but no comparison was made between the predictions of their model and experimental data.

A more quantitative connection between earthquake statistics and percolation was made by Sahimi *et al.* (1993) who investigated the fractal properties of the *spatial distribution* of earthquake hypocenters. These authors analyzed four seismic data sets from four different regions in California. For each set, the fractal dimension D_f of the set was estimated by counting the minimum number of cubes N_c of edge length L to completely cover the three-dimensional data set at various length scales L, where L was taken to be location error of the set. Figure 4.5 presents their results, from which $D_f \simeq 1.8$ for *all* cases.

What is the interpretation of this value of D_f? Geologic features are rarely simple planes. Most faults are comprised of many shear fractures, which are visible at a number of scales. Therefore, given that fracture networks of rock at large scales appear to be the largest percolation clusters at p_c, the fault patterns should also be percolation clusters with $D_f \simeq 2.5$ in three dimensions. The vast majority of earthquakes are distributed on the regional fault networks. But the hypocenters have to belong to the active part of the network, i.e., the part where finite strain deformation can take place. The easiest way for this to happen is if the hypocenters belong to the *backbone* of the fault network. Indeed, the result $D_f \simeq 1.8$ is in complete agreement with the fractal dimension $D_{BB} \simeq 1.8$ of three-dimensional percolation backbone (see Table 2.3). Moreover, the fact that all four data sets yielded the same D_f is consistent with the universality of D_{BB}.

Kagan and Knopoff (1981) studied the spatial distribution of earthquakes but found $D_f \simeq 2$, which they interpreted as meaning that earthquakes occur on *planes*. But, they also found that if they explore larger depths and areas, the estimated D_f would decrease. If the hypocenters considered by Sahimi *et al.* (1993) are plotted in three-dimensional space and rotated around all three axes, there does not seem to be any one plane about which the earthquakes cluster. The San Andreas–Elsinore (SA–EL) earthquakes are located between two major faults and over a very wide zone of deformation, whereas the Parkfield earthquakes are located over a relatively thin zone of deformation along the SA faults, yet they both have $D_f \simeq 1.8$. There usually exists a large length-scale cutoff in the three-dimensional distribution of earthquake hypocenters, namely, the brittle crust length, beyond which we may expect a crossover to a quasi-two-dimensional distribution. However, even in the SA–EL case, where the earthquakes are distributed over a very wide region, such a crossover is not evident. As a further check of this hypothesis, Nakanishi *et al.* (1993) studied dynamical properties, e.g., diffusion properties, of the clusters formed by earthquake hypocenters. They found them to be completely consistent with those of percolation backbones, thus providing further support that earthquakes, fault and fracture patterns, and percolation are closely related phenomena.

For an alternative view of earthquakes and their dynamical properties see Bak and Tang (1989) and Sornette and Sornette (1989)

4.6 Conclusions

Fracture and fault patterns in macroscopically heterogeneous rock may be similar to percolation networks. If so, the vast knowledge about percolation networks can be used for modeling fracture networks. This in turn would facilitate modeling flow phenomena in fractured rocks, an important but unsolved problem. Earthquakes may also have close connections with percolation which would help us better understand this complex phenomenon.

References

Bak, P. and Tang, C., 1987, *J. Geophys. Res.* **B94**, 15635.

Barton, C.C. and Hsieh, P.A., 1989, *Physical and Hydrological-Flow Properties of Fractures*, Guidebook T385, Las Vegas, American Geophysical Union.

Chelidze, T., 1981, *Doklady Akad. Nauk SSSR* **246**, 14.

Chelidze, T. and Gueguen, Y., 1990, *Int. J. Rock. Mech. Min. Sci. Geomech. Abtr.* **27**, 223

de Arcangelis, L., Hansen, A., Herrmann, H.J., and Roux, S., 1989, *Phys. Rev. B* **40**, 877.

Englman, R., Gur, T. and Jaeger, Z., 1983, *J. Appl. Mech.* **50**, 707.

Gutenberg, B. and Richter, C.F., 1956, *Ann. Geofis.* **9**, 1.

Haldorsen, H.H., Brand, P.J. and MacDonald, C.J., 1988, in *Mathematics of Oil Production*, edited by S.F. Edwards and P.R. King, Oxford, Clarendon Press.
Herrmann, H.J. and Roux, S. (eds.), 1990, *Statistical Models for the Fracture of Disordered Media*, Amsterdam, North Holland.
Hestir, K., and Long, J.C.S., 1990, *J. Geophys. Res.* **B95**, 21565.
Hewett, T.A. and Behrens, R.A., 1990, *SPE Formation Evaluation* **5**, 217.
Hirata, T., 1989, *Pure Appl. Geophys.* **131**, 157.
Kagan, Y.Y. and Knopoff, L., 1981, *J. Geophys. Res.* **86**, 2583.
Kantor, Y. and Webman, I., 1984, *Phys. Rev. Lett.* **52**, 1891.
Kirkpatrick, S., Gelatt, C.D., and Vecchi, M.P., 1983, *Science* **220**, 671.
Long, J.C.S., Karasaki, K., Davey, A., Landsfeld, M., Kemeny, J., and Martel, S., 1991, *Int. J. Rock Mech. Min. Sci. Geomech. Abstr.* **28**, 121.
Nakanishi, H., Sahimi, M., Robertson, M.C., Sammis, C.G., and Rintoul, M.D., 1993, *J. Phys. France I* **3**, 773.
Okubo, P.G. and Aki, K., 1987, *J. Geophys. Res.* **92**, 345.
Otsuka, M., 1979, *Butsuri* **34**, 509.
Sahimi, M., 1992, *Physica A* **186**, 160.
Sahimi, M., 1993, *Rev. Mod. Phys.*
Sahimi, M. and Arbabi, S., 1992, *Phys. Rev. Lett.* **68**, 608.
Sahimi, M. and Goddard, J.D., 1986, *Phys. Rev. B* **33**, 7848.
Sahimi, M., Robertson, M.C., and Sammis, C.G., 1993, *Phys. Rev. Lett.* **70**, 2186.
Sornette, A. and Sornette, D., 1989, *Europhys. Lett.* **9**, L197.
Stark, C.P. and Stark, J.A., 1991, *J. Geophys. Res.* **B96**, 8417.
Trifu, C.-I. and Radulian, M., 1989, *Phys. Earth Plant. Interiors* **58**, 277.
Wu, R.-S. and Aki, K., 1985, *Pure Appl. Geophys.* **123**, 806.

5
Single-phase flow and transport in porous media

5.0 Introduction

In this chapter we discuss the application of percolation theory to transport and flow of one fluid in porous media and fractured rocks. We focus on low Reynolds number (low velocity) flow, where Darcy's law is applicable. In Chapter 3 we learnt that porous media can often be represented by a network whose sites and bonds represent, respectively, the pore bodies and pore throats of a pore space. Transport properties of such networks are controlled by the throats, which is why the role of pore bodies is usually neglected (no volume is assigned to the sites). Chapter 4 told us that large-scale fracture networks may be nothing but percolation clusters. Thus, we can develop a unified approach to tranport and flow in both porous media and fractured rock, and use various percolation-based methods for estimating the permeability k and electrical conductivity σ of fluid-saturated rock (the conductivity of the solid matrix is ignored). Of course, as a result of Einstein's relation (Chapter 2), the electrical conductivity is proportional to the effective diffusivity of the system. Every method discussed in this chapter can also be used for calculating the effective properties of disordered materials such as polymers, glasses, powders, and metal films, and in fact the literature on this subject is very large. This class of applications of percolation concepts is discussed in Chapter 12.

5.1 Exact results and rigorous bounds

It has not yet been possible to calculate k and σ exactly for a porous medium of *arbitrary* microstructure. The available exact results are for a few models of porous media with simple microstructure, in which no disorder is allowed and, moreover, most of the results are restricted to high porosities. Even then most of the exact results are for periodic arrays of spherical particles, in which the spheres have the same radius R and are placed at the nodes of a regular lattice. These models are idealizations of *unconsolidated* porous media. Hasimoto (1959) was the first to treat slow fluid flow through a dilute cubic array of spheres. He derived the periodic fundamental solution to the Stokes' equation

$$-\nabla P+\rho\mathbf{g}+\eta\nabla^2\mathbf{v}=0, \tag{5.1}$$

to derive an expression for k. Here P is the pressure, ρ, η and $\mathbf{v}$ are, respectively, the density, viscosity, and the average velocity of the fluid, and $\mathbf{g}$ is the gravity vector. Since Hasimoto's paper, many other authors have studied this problem. In particular, Sangani and Acrivos (1982) obtained expressions for the permeability of various lattice models of spheres. For example, if c is the volume fraction of the spheres, then their result for a simple cubic lattice of spheres is given by

$$\frac{k_s}{k}=1-1.7601c^{1/3}+c-1.5593c^2+3.9799c^{8/3}-3.0734c^{10/3}+O(c^{11/3}), \tag{5.2}$$

where $k_s=2R^2/(9c)$ is called the Stokes' permeability. Sangani and Acrivos (1982) also presented the numerical coefficients in the above expansions for body-centered cubic (BCC) and face-centered cubic (FCC) lattices, and two-dimensional square and hexagonal lattices of parallel cylinders. They also showed that these expansions are convergent for $0<c/c_{max}<0.85$, where c_{max} is the maximum volume fraction of the spheres for a given packing and, $c_{max}=\pi/6$, $3^{1/2}\pi/8$, and $2^{1/2}\pi/6$ for simple cubic, BCC, and FCC lattices, respectively. Since the spheres or the cylinders do not overlap, these results may be relevant to unconsolidated porous media. On the other hand, if we consider flow through a *random* array of spheres, the following asymptotic expression for k is obtained (Hinch 1977, Kim and Russel 1985)

$$\frac{k_s}{k}=1+\frac{3}{\sqrt{2}}c^{1/2}+\frac{135}{64}c\log c+16.456c\ldots \tag{5.3}$$

For a more complete discussion of these results see Sahimi (1993).

Another set of rigorous results for random porous media is obtained when, instead of trying to solve the problem exactly, we derive upper and lower bounds to the properties of interest. These bounding methods have been reviewed by Torquato (1991). For example, one can obtain upper and lower bounds to the permeability of the Swiss cheese model discussed in Chapter 2, or a model in which the spheres have a specific degree of impenetrability (remember that in the Swiss cheese model the spheres have an arbitrary degree of impenetrability). The degree of impenetrability affects the percolation threshold c of the particle (sphere) phase. We have $c = 0.34$ for the Swiss cheese model and $c = 0.64$ for the totally impenetrable sphere model.

In all the above studies, the Kozeny–Carman empirical formula

$$\frac{k_s}{k} = \frac{10c}{(1-c)^3}, \tag{5.4}$$

(recall that $1 - c = \phi$ is the porosity) falls within 15% of the results for at least one of the three types of periodic packings if $c > 0.5$, and in the case of the random sphere model it is relatively close to the lower bound.

The main problem with all of these results is that, they are not very useful for the highly disordered (and consolidated) porous media that are of interest to us in this book. If $\phi = 1 - c$ is close to the percolation threshold, then all of the above formulae break down. In fact, no bound, no matter how accurate, can predict the existence of a nontrivial (not zero or unity) percolation threshold.

5.2 Effective-medium approximation

Effective-medium approximation (EMA) is a method of determining the effective properties of a disordered medium, in which the medium is replaced with a hypothetical uniform medium with unknown properties. An excellent review of this subject is given by Landauer (1978). In its original form, the EMA had nothing to do with percolation. But it was the first analytical approximation able to predict a nontrivial percolation threshold and, for this reason, it has been closely tied to percolation. There are two typical EMAs, but the one that is of interest to us was originally developed by Bruggeman (1935) and Landauer (1952). In this approach, the disordered medium is replaced by a uniform system of unknown properties, which is supposed to mimic the disordered system. The distribution of the field (pressure, voltage, stress, etc.) in this uniform medium is also uniform, and can be calculated analytically. An inclusion of the original system, which in principle can have any shape, but in practice is either spherical or ellipsoidal (or is a bond in the network models; see below), is then inserted in the uniform medium. This insertion induces an extra field in the uniform medium. We insist that the average of this extra field, when the averaging is taken with respect to the distribution of disorder in the original system, must be zero. This *self-consistent condition* yields an equation for the unknown property of the uniform system, which is also an estimate for the property of the disordered medium.

Kirkpatrick (1973) extended the Bruggeman–Landauer EMA to percolation networks of bonds with random conductances. Consider a regular network of coordination number Z and pore conductance distribution

$h(g_p)$. Then, EMA predicts that g_e, the effective conductance of the network is the solution of the following equation

$$\int_0^\infty h(g_p)\frac{g_p - g_e}{g_p + (\gamma^{-1} - 1)g_e}\,dg_p = 0, \tag{5.5}$$

where $\gamma = 2/Z$. The same equation was derived by Bruggeman and Landauer with $\gamma^{-1} = d$ for a d-dimensional continuous system. Using this equation, we calculate the permeability and conductivity of a porous medium. Consider an effective-medium network where each bond or pore has a conductance g_e. We fix the pressures at two opposite faces of the network so as to produce an average pressure gradient $\langle \nabla P \rangle$. The total fluid flux q crossing any plane perpendicular to $\langle \nabla P \rangle$ is the sum of the individual fluxes in the bonds intersecting the plane. Each flux is the pressure difference across the bond times g_e/η. If we approximate the local pressure difference as the projection of the average pressure gradient along the bond length $\mathbf{l}$, we find

$$q = \sum \frac{g_e}{\eta}\langle \nabla P \rangle \cdot \mathbf{l}. \tag{5.6}$$

If we divide q by the area S of the plane, we obtain an average velocity which, when compared with Darcy's law, yields an estimate of the permeability

$$k = g_e\left\langle \frac{1}{S}\sum \mathbf{l}\cdot\mathbf{n}\right\rangle, \tag{5.7}$$

where $\mathbf{n}$ is a unit vector along the pressure gradient. But, if the medium is statistically homogeneous and isotropic, any unit vector can be used. Equation (5.7) shows clearly why, for random networks, k and g_e obey the same scaling law near the percolation threshold (see Chapter 2). In a similar way, the electrical conductivity σ of a fluid-saturated network, which is simply proportional to g_e, can be calculated.

In this kind of approach, we neglect the pressure drop in a pore body and we assume that most of the pressure drop occurs in the pore throats of the network. We do not do the case in which both pore bodies and pore throats are considered. So long as pore bodies are large and compact and pore throats are long and narrow, the approximation is valid.

A stringent test for any theory of transport in porous media is its ability to predict the percolation threshold of the system. Despite its simplicity the EMA can predict a nontrivial percolation threshold for various networks. If we use, $h(g_p) = (1 - p)\delta(g_p) + pf(g_p)$, i.e., a network in which a fraction $(1 - p)$ of the pores are nonconducting (do not allow any fluid flow), and the conductance of the rest is selected from the distribution $f(g_p)$, then EMA predicts that g_e vanishes at $p_c = 2/Z$. This prediction is accurate for two-dimensional networks, but not for three-dimensional networks (see

Tables 2.1 and 2.2). If we take $f(g_p) = \delta(g_p - g_0)$, i.e., all conducting bonds have the same conductance g_0 with probability p, then (5.5) predicts that

$$\frac{g_e}{g_0} = \frac{p - 2/Z}{1 - 2/Z} = \frac{p - p_c}{1 - p_c}, \tag{5.8}$$

i.e., a straight line between $p = 1$ and $p = 2/Z = p_c$. Thus, EMA predicts that the conductivity critical exponent μ is one in *all* dimensions, which is a wrong result (see Table 2.3). In general, EMA is very accurate if the system is not close to p_c, *regardless* of the structure of $h(g_p)$. Its predictions are also more accurate for two-dimensional networks than for three-dimensional ones.

5.2.1 Effective-medium approximation for transient diffusion and conduction

Equation (5.5) is valid for steady state transport, but can be extended to time-dependent diffusion and conduction processes. The time-dependent transport equation is given by

$$\frac{\partial P_i}{\partial t} = \sum_j W_{ij}[P_j(t) - P_i(t)], \tag{5.9}$$

which can be viewed as a description of the motion of a diffusing particle on a lattice. Equation (5.9) is usually called the *master equation*. Here $P_i(t)$ is the probability that the particle will be at site i at time t, and W_{ij} is the transition rate (probability) between sites i and j, taken to be nonzero only if sites i and j are nearest neighbors. The transition rates are randomly distributed variables and are usually independent of time (see dynamic percolation in Chapter 1 for the case in which W_{ij} is time dependent). To derive an EMA for this problem we have to treat it in the Laplace transform space. Then, it can be shown that EMA predicts that (Odagaki and Lax 1981, Webman 1981, Sahimi *et al.* 1983a)

$$\int_0^\infty h(w) \frac{1}{1 - p_c + \varepsilon p_c G_0(\varepsilon) + p_c[1 - \varepsilon G_0(\varepsilon)](w/W_m)} dw = 1. \tag{5.10}$$

Here, $p_c = 2/Z$, $\varepsilon = \lambda/W_m$, where λ is the Laplace transform variable conjugate to the time, W_m is the Laplace transform of the effective (time-dependent) transport coefficient, thus, $g_e = W_m(0)$, and $G_0(\varepsilon)$ is a Green function which, for a d-dimensional simple cubic network, is given by

$$G_0(\varepsilon) = \int_0^\infty \exp\left[-\frac{1}{2}x(2d + \varepsilon)\right] I_0^d(x)\, dx, \tag{5.11}$$

where $I_0(x)$ is the modified Bessel function of order zero. For a table of the Green functions for other networks see Sahimi *et al.* (1983a). In the limit, $\lambda = 0$ (i.e., very long times), (5.10) reduces to (5.5). If we set $\lambda = i\omega$, where $i = \sqrt{-1}$ and ω is the frequency, then (5.10) can be used for estimating frequency-dependent transport properties (i.e., AC properties) of disordered media, discussed in Chapter 12.

5.2.2 Effective-medium approximation for anisotropic systems

Many disordered media such as fracture networks are *anisotropic*, and therefore are characterized by a permeability or conductivity *tensor*. Thus, the effective network should be characterized by three effective conductances g_{e1}, g_{e2}, and g_{e3}, one for each principal direction of the network. Each principal direction is also characterized by its own conductance distribution. Bernasconi (1974) developed an EMA for anisotropic media which, for a d-dimensional simple cubic network, is given by

$$\int_0^\infty h_i(x)\,\frac{g_{ei} - x}{x + S_i}\,dx = 0, \quad i = 1, \ldots, d, \tag{5.12}$$

where h_i is the conductance distribution in the ith direction. For a square network

$$S_1 = g_{e1}\,\frac{\arctan(g_{e2}/g_{e1})^{1/2}}{\arctan(g_{e1}/g_{e2})^{1/2}}, \tag{5.13}$$

and S_2 is obtained by interchanging g_{e1} and g_{e2}. For a simple cubic network

$$S_1 = g_{e1}\,\frac{\arctan[g_{e1}^{-1}(g_{e1}g_{e2} + g_{e1}g_{e3} + g_{e2}g_{e3})]^{1/2}}{\arctan[g_{e1}(g_{e1}g_{e2} + g_{e1}g_{e3} + g_{e2}g_{e3})]^{-1/2}}, \tag{5.14}$$

and S_2 and S_3 are obtained by cyclic rotation of g_{e1}, g_{e2}, and g_{e3}. These equations allow one to calculate the permeability tensor of a fracture network. Hereafter, we refer to (5.5) and (5.12) as the isotropic and anisotropic EMA, respectively. We take, for example, $h_1(g_p) = p\delta(g_p - 10) + (1 - p)\delta(g_p)$, and $h_2(g_p) = p\delta(g_p - 1) + (1 - p)\delta(g_p)$. The results for g_{e1} and g_{e2} for the square network are shown in Fig. 5.1. They show very good agreement with Monte Carlo simulation results. Unlike the isotropic case, the results do not follow straight lines.

Koplik *et al.* (1984) analyzed in detail a Massilon sandstone, mapped its pore space onto an equivalent random network, and employed the isotropic EMA to calculate its permeability and conductivity. They found that the predictions for k differ from the data by about one order of magnitude, while those for σ differ by a factor of about 2. They attributed the difference

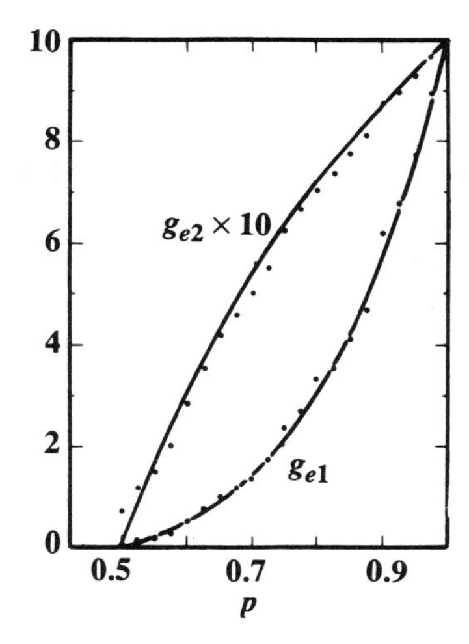

Figure 5.1 *Effective conductivities g_{e1} and g_{e2} for the anisotropic bond percolation on the square network. The curves represent the EMA predictions; circles are Monte Carlo data (after Bernasconi 1974).*

to the fact that most sedimentary rocks are highly heterogeneous and anisotropic, none of which is adequately taken into account by the isotropic EMA. Doyen (1988) analyzed transport properties of Fontainebleau sandstones and used the isotropic EMA to predict their k and σ, and found that both of them can be predicted to within a factor of 3. Benzoni and Chang (1984) used the isotropic EMA to estimate the effective diffusivity of high-area alumina catalyst particles. Their results are shown in Fig. 5.2 and, in this case, the agreement between the predictions and the experimental data is excellent. Harris (1992) used the anisotropic EMA to calculate the permeability tensor of heterogeneous and fractured rock. The application of EMA to predicting transport properties of other kinds of disordered media is discussed in Chapter 12.

5.2.3 Archie's law

A useful empiricism for sedimentary rocks is Archie's law given by

$$\sigma = \sigma_f \phi^m, \tag{5.15}$$

where σ_f is the conductivity of the fluid saturating the porous medium (usually brine). The exponent m has been found to vary anywhere between 1.3 and 4, depending upon consolidation and other factors. Archie's law has been found to hold for a wide variety of rocks. It implies that the fluid phase remains connected and its percolation threshold is always zero. Percolation concepts and EMA have been used by Sen *et al.* (1981) and

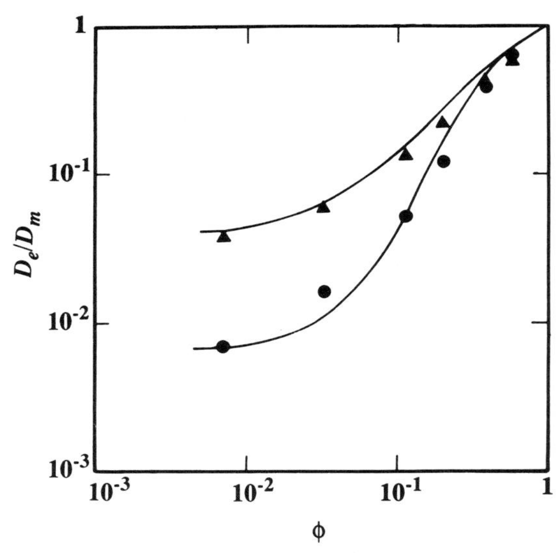

Figure 5.2 *Comparison of the EMA predictions of the dependence of the effective diffusivity D_e of alumina catalyst particles with high surface area on the porosity φ (curves) with the experimental data. The porous medium is modeled by a simple cubic network of macropores and micropores. The upper curve represents the case $a = 0.025$, where a is the ratio of micro and macro diffusivities; the lower curve is for $a = 0.004$. D_m is the molecular diffusivity. Triangles and circles are the data at pressures 1 atm and 10 atm, respectively (after Benzoni and Chang 1984).*

Mendelson and Cohen (1982) for deriving Archie's law, showing that, $m = Z/(Z - 2)$. Thus, consistent with experimental observations, m is not universal.

One drawback of an EMA-based derivation of (5.15) is that it pertains only to a microstructure whose solid component is disjoint, whereas in reality the solid matrix is continuous and sample-spanning. This difficulty can be circumvented by using an EMA for a *three-component* system consisting of fluid, solid and *cement material*. This was done by Sheng (1990) who also showed that such an EMA reproduces Archie's law with $m = (5 - 3L)/[3(1 - L^2)]$, where L is the depolarization coefficient of the grains. For a more complete discussion of this important subject see Sahimi (1993).

Although EMA has been successful in providing a derivation of Archie's law, its use for understanding the behavior of rocks is not without conceptual difficulty. Generally speaking, rocks have porosities less than 40%. This is far from the regime in which the assumptions, based on which (5.15) is derived, can be justified. If the porosity of the pore space is low, then the grains are in close contact with one another and the interaction between them is important. Such interactions cannot be taken into account by EMA. Moreover, rocks with very similar grains can have very different values of m, and rocks with very dissimilar grains can have very similar values of m. These cannot be explained with such EMAs.

5.3 Position–space renormalization group methods

The main assumption behind any EMA is that the mean fluctuations in the potential field in the system, and thus in the transport properties are small. However, if the fluctuations are large as in, e.g., a fractal porous medium, or one that is near its p_c, or in a macroscopically heterogeneous medium with a broad distribution of the permeabilities, EMA breaks down and loses its accuracy. In such cases, a position–space renormalization group (PSRG) method is more accurate. We describe a PSRG method for a random network model of pore space. Its generalizations to other more complex systems are reviewed by Stanley *et al.* (1982).

For a percolating network, a PSRG method can be used for simultaneous prediction of the percolation threshold p_c, the correlation length exponent ν_p, and the conductance g_e and its critical exponent μ. Calculating all of these quantities by the same method is a distinct advantage of PSRG methods. Consider, for example, a square or a cubic network in which each bond is conducting with probability p. The idea in any PSRG method is that, since our network is so large that we cannot calculate its properties exactly, we partition it into $b \times b$ or $b \times b \times b$ cells, where b is the number of bonds in any direction, and calculate their properties, which are hopefully representative of the properties of our network. The shape of the cells can be selected arbitrarily, but it should be chosen in such a way that it preserves, as much as possible, the properties of the network. For example, an important topological property of the square network is that it is *self-dual*. The dual of a network is obtained by connecting the centers of the neighbouring polygons (or polyhedra) that constitute the network. For example, if we connect the centers of the hexagons in a hexagonal network, we obtain the triangular network, and vice versa. Thus, these networks are duals of each other. However, if we connect the centers of the squares in a square network, we again obtain a square network, and thus this network is self-dual. This self-duality plays an important role in the percolation properties of the square network, and therefore it would be desirable to partition it into cells which are also self-dual. Figure 5.3 shows an example of such $b = 2$ cells for the square and cubic networks. The two-dimensional cell is self-dual.

The next step in a PSRG method is to replace each cell with one bond in each principal direction. If in the original network each bond is conducting with probability p, then, the bonds that replace the cells would be conducting with probability $p' = R(p)$; this is also shown in Fig. 5.3 $R(p)$ is called the *renormalization group transformation*, and is the *sum* of the probabilities of *all* conducting configurations of the RG cell, obtained as follows. Since

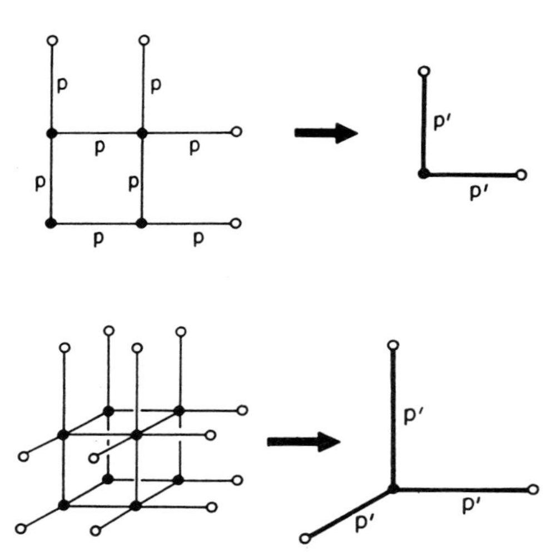

Figure 5.3 *Two- and three-dimensional RG cells (left) and their renormalized equivalents (right).*

we are interested in percolation and transport in our network, and because the RG cells are supposed to represent the network, we solve the percolation and transport problem in each cell by applying a fixed potential difference across the cell in a given direction. For example, consider the 2×2 cell of Fig. 5.3, displayed again in Fig. 5.4. Since we are interested in diffusion and flow, which are linear problems, the magnitude of the potential difference across the cell is not important. Thus, we hold sites A

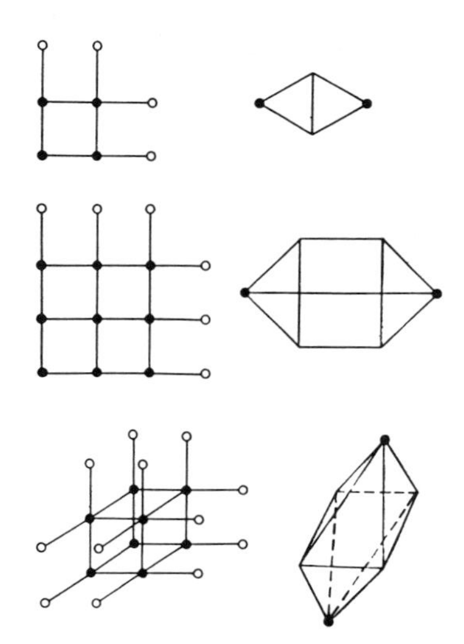

Figure 5.4 *Two- and three-dimensional RG cells (left) and their equivalent electrical circuits (right).*

and B (on the left) at a unit potential, and sites C and D (on the right) at zero potential. Since A and B, and C and D have the same potential they are equivalent, and therefore can be reduced to just two nodes A and C. This means that as far as percolation and transport are concerned, the 2×2 cell of Fig. 5.4 is equivalent to the circuit shown there, which is usually called the Wheatstone bridge. Thus, for this cell we only need to deal with five bonds, for the 3×3 cell we only have to worry about the 13-bond circuit, and for the $2 \times 2 \times 2$ cell we have an equivalent 12-bond circuit (see Fig. 5.4). To obtain $R(p)$, we find all configurations of such circuits, with some bond conducting and some insulating (missing), such that there is a transport path from left to right (or from top to bottom), or if we start walking along the conducting bonds from the left, we can find a path to reach the opposite, which is also the usual definition for percolation. Thus, for the 2×2 cell we obtain

$$p' = R(p) = p^5 + 5p^4q + 8p^3q^2 + 2p^2q^3, \tag{5.16}$$

whereas the 3×3 cell yields

$$p' = p^{13} + 13p^{12}q + 78p^{11}q^2 + 283p^{10}q^3 + 677p^9q^4 + 1078p^8q^5 + 1089p^7q^6 + 627p^6q^7 + 209p^5q^8 + 38p^4q^9 + 3p^3q^{10}, \tag{5.17}$$

and for the $2 \times 2 \times 2$ cell we have

$$p' = p^{12} + 12p^{11}q + 66p^{10}q^2 + 220p^9q^3 + 493p^8q^4 + 776p^7q^5 + 856p^6q^6 + 616p^5q^7 + 238p^4q^8 + 48p^3q^9 + 4p^2q^{10}, \tag{5.18}$$

where $q = 1 - p$. To see how these equations are obtained, consider (5.16) for the 2×2 cell. There is only one conducting cell configuration with all five bonds conducting (probability p^5), five conducting configurations with four bonds conducting and one bond insulating (probability $5p^4q$), eight conducting configurations with three bonds conducting and two bonds insulating (probability $8p^3q^2$), and only two conducting configurations with two bonds conducting and three bonds insulating (probability $2p^2q^3$).

At p_c the sample-spanning cluster is self-similar at all length scales (see Chapter 2). Thus, partitioning the network into cells, and replacing them with bonds should not change the properties of the network at p_c, because the network should look the same at p_c at any length scale, and replacing the cells with bonds only changes the length scale. This means that the RG transformation should remain invariant at p_c. The same thing should be true at $p = 1$ and $p = 0$, i.e., under any transformation full and empty networks should be transformed to full and empty networks again. The points $p = p_c$, 0, 1 are called the *fixed points* of the network. Since the RG transformation should not change anything at these points, the implication is that at these points the probability of having a conducting bond in the original network p and the probability of the bonds that replace the cells $[p' = R(p)]$ should be the same. Thus, the fixed points should be the solution of the equation

$$p = R(p). \tag{5.19}$$

Indeed this equation has three roots: $p = 0$, $p = 1$, and $p = p^*$, where p^* is the RG prediction for p_c, which may or may not be the same as the true value of p_c. For the RG cells of Fig. 5.4 we obtain $p^* = 1/2$ for both 2×2 and 3×3 cells, which is also the exact bond percolation threshold of the square network. In fact, it can be shown that the RG transformations for such cells for *any* b would always predict the exact value $p_c = 1/2$. For the $2 \times 2 \times 2$ cell we obtain $p^* \simeq 0.208$, which should be compared with the numerical estimate $p_c \simeq 0.2488$.

Each time the RG cell is replaced with one bond in each direction, the correlation length is reduced by a factor b, since the length of the bonds replacing the cells is b times larger than those in the original network. That is, $\xi_p = b\xi_p'$, where ξ_p' is the correlation length in the network built up of the bonds that have replaced the cells. In the original network the correlation length follows the scaling law, $\xi_p \sim (p - p_c)^{-\nu_p}$ near p_c, and in the new network we have $\xi_p' \sim (p' - p^*)^{-\nu_p} \sim [R(p) - R(p^*)]^{-\nu_p}$. Thus, if we linearize $R(p)$ about $p_c(p^*)$, then it is not difficult to see that

$$\nu_p = \frac{\ln b}{\ln \lambda_p}, \tag{5.20}$$

where $\lambda_p = dR(p)/dp$, evaluated at p^*. Thus, for the 2×2 and 3×3 cells we obtain $\nu_p \simeq 1.43$ and 1.38, respectively, which should be compared with the exact value $\nu_p = 4/3$. For the $2 \times 2 \times 2$ cell we get $\nu_p \simeq 1.03$, which should be compared with the true value $\nu_p \simeq 0.88$. As b increases, the estimate of ν_p approaches its true value.

We can now discuss the PSRG approach for conductivity and permeability of percolation networks. Suppose that the distribution of the pore conductance (or permeability) of the network is $h_0(g_p)$. Since we partition our network into cells and replace them with one bond in each principal direction, the conductance distribution of these bonds is no longer $h_0(g_p)$ because, as discussed above, these bonds are conducting with probability $p' = R(p)$, which is the sum of the probabilities of all conducting configurations of the cell. Suppose that the conductance distribution of these bonds is $h_1(g_p)$. This new distribution is related to $h_0(g_p)$ through the following equation

$$h_1(g_p) = \int h_0(g_1)dg_1 h_0(g_2)dg_2 \ldots h_0^{*}(g_n)dg_n \delta(g_p - g'), \tag{5.21}$$

where $g_1, \ldots, g_n$ are the conductances of the n bonds of the RG cell, and g' is the equivalent conductance of the bond replacing it. For example, for the 2×2 cell of Fig. 5.4 we have

$$g' = \frac{g_1(g_2g_3 + g_2g_4 + g_3g_4) + g_5(g_1 + g_2)(g_3 + g_4)}{(g_1 + g_4)(g_2 + g_3) + g_5(g_1 + g_2 + g_3 + g_4)}. \tag{5.22}$$

If $h_0(g_p) = (1 - p)\delta(g_p) + p\delta(g_p - g_0)$, then (5.21) becomes very simple and yields

$$h_1(g_p) = [1 - R(p)]\delta(g_p) + \Sigma_i a_i(p)\delta(g_p - g_i), \tag{5.23}$$

where g_i is the conductance of a conducting configuration of the cell that occurs with probability $a_i(p)$, and the sum is over all conducting cell configurations. The meaning of the first term on the right side of (5.23) is clear: Since $R(p)$ is the probability that the bonds replacing the cells are conducting, $1 - R(p)$ is the probability that they are insulating. To understand how the various terms of the sum in (5.23) arise, consider those conducting configurations of the 2×2 cell of Fig. 5.4 in which three bonds are conducting and two bonds are insulating. As (5.16) indicates, there are eight such configurations. The conductance of two of them is $g_0/3$, hence $a_i(p) = 2p^3q^2$, and $g_i = g_0/3$. The conductance of the other six configurations is $g_0/2$, and therefore $a_i(p) = 6p^3q^2$ and $g_i = g_0/2$. In this manner, the conductance of all conducting configurations and their corresponding probabilities can be calculated. It may be obvious that, $\Sigma_i a_i(p) = R(p)$. Thus, for the 2×2 cell we find

$$h_1(g_p) = [1 - R(p)]\delta(g_p) + 2p^3q^2\delta\left(g_p - \frac{1}{3}g_0\right) + 2p^2(1 + 2p)q^2\delta\left(g_p - \frac{1}{2}g_0\right)$$
$$+ 4p^3q\delta\left(g_p - \frac{3}{5}g_0\right) + p^4\delta(g_p - g_0), \tag{5.24}$$

which is already more complex than $h_0(g_p)$.

We now partition our new network with pore conductance distribution $h_1(g_p)$ into $b \times b$ cells, and replace each cell with one bond in each principal direction. The conductance distribution of these new bonds is $h_2(g_p)$. This new distribution will also have the same form as (5.23), except that the sum term has many more terms than that for $h_1(g_p)$. If this iteration continues, we will finally obtain a distribution $h_\infty(g_p)$ whose shape does not change under further interations. In analogy with the fixed points of the RG transformation, we call this the *fixed-point distribution*. The conductivity of our original network would be the average of this distribution. But it is difficult to carry on this iteration process analytically more than once or twice, because the number of δ-functions in (5.23) increases very rapidly and, as a result, calculating $h_\infty(g_p)$ analytically is not practical. The common practice is to replace $h_i(g_p)$ at the ith iteration by an *optimum* distribution $h_{io}(g_p)$ such that it mimics the properties of $h_i(g_p)$ closely. Many schemes for calculating this optimum distribution have been proposed. One of the most accurate approximations was proposed by Bernasconi (1978). In this scheme the optimum distribution $h_{1o}(g_p)$ for $h_1(g_p)$ has the following form

$$h_{1o}(g_p) = [1 - R(p)]\delta(g_p) + R(p)\delta[g_p - g_o(p)], \tag{5.25}$$

where $g_o(p)$ is given by

$$g_o(p) = \exp\left[\frac{1}{R(p)} \Sigma_i a_i(p) \ln g_i\right], \tag{5.26}$$

and $a_i(p)$ and g_i are the same as those in (5.23). Once $g_o(p)$ is calculated, (5.21) is iterated again, a new distribution $h_2(g_p)$ and its optimum $h_{2o}(g_p)$ are determined, and so on. In practice, after n iterations, where $n = 3$–5, even a broad $h_0(g_p)$ converges to the fixed-point distribution. Then, the conductance $g_o(p)$, obtained after n iterations, turns out to be a very good approximation to g_e. Figure 5.5 compares $g_o(p)$ with Monte Carlo data for g_e in the square network, and the agreement is very good. The same method can be used for estimating the permeability of the network. Moreover, one

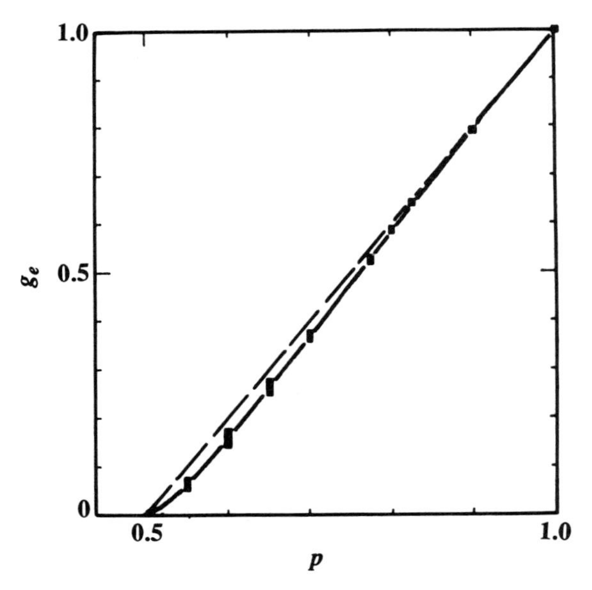

Figure 5.5 *Comparison of the PSRG predictions of the conductivity g_e of the square network (curve) with the Monte Carlo data (rectangles), as the fraction p of the conducting bonds is varied. The upper dashed line represents the EMA predictions (after Bernasconi 1978).*

finds after n interations that $\lambda_c h_{n+1}(\lambda_c g_p) \simeq h_n(g_p)$, where λ_c is a constant. The critical exponent μ is then calculated from

$$\mu = \nu_p \frac{\ln \lambda_c}{\ln b}. \tag{5.27}$$

For example, for both 2×2 and 3×3 cells of Fig. 5.4 we find $\mu \simeq 1.32$, in excellent agreement with the accepted value, $\mu \simeq 1.3$. For the $2 \times 2 \times 2$ cell we find $\mu \simeq 2.14$, about 7% larger than the accepted value $\mu \simeq 2$ (see Table 2.3).

PSRG methods are very accurate for two-dimensional systems (see Fig. 5.5). They can also be used for anisotropic media such as fracture networks. In this case we have to derive d RG transformations for a d-dimensional system, and (5.21) should also be used for each principal direction of the system. PSRG methods do have a major drawback for three-dimensional systems – the predicted p_c with a $2 \times 2 \times 2$ cell of *any* shape is not very accurate. This means that g_e and k would not be predicted accurately near p_c. Therefore, we have to use larger RG cells. However, even for a $b = 3$ cell, the RG transformation is difficult to calculate analytically, because the number of possible configurations is of the order of 3×10^{11}. Thus, we may have to resort to a Monte Carlo sampling for constructing $R(p)$ and $h_1(g_p)$, in which case the method would not be any simpler than a direct Monte Carlo calculation of the effective conductivity or permeability.

5.4 Renormalized effective-medium approximation

Sections 5.2 and 5.3 show that EMA and PSRG methods have their strengths and their weaknesses. Sahimi *et al.* (1983b) proposed another method, called the renormalized EMA (REMA), which combines EMA and PSRG methods to take advantage of their strengths and to eliminate their shortcomings. The idea is that, EMA is very accurate away from p_c, and under a RG transformation the correlation length reduces ($\xi_p' = \xi_p/b$), and thus the renormalized network is farther away from p_c. Hence, we may apply EMA to the renormalized network, instead of the original one. That is, the pore conductance distribution that we use in (5.5) is $h_1(g_p)$ instead of $h_0(g_p)$. This method increases drastically the accuracy of the results. For example, with the $2 \times 2 \times 2$ cell of Fig. 5.3, we obtain, $p_c^* \simeq 0.265$, only 7% larger than $p_c \simeq 0.2488$ for the cubic network. In general, REMA predicts that p_c is the root of $R(p_c) = 2/Z$ (instead of $p_c = 2/Z$, predicted by EMA, and $R(p_c) = p_c$, the PSRG estimate). Using this method, Sahimi *et al.* (1983b) obtained accurate predictions of conductivity of various networks. Figure 5.6 compares the prediction of the REMA with the Monte Carlo data for the conductivity of the cubic network, and the predictions of EMA and the PSRG method. REMA provides the most accurate predictions.

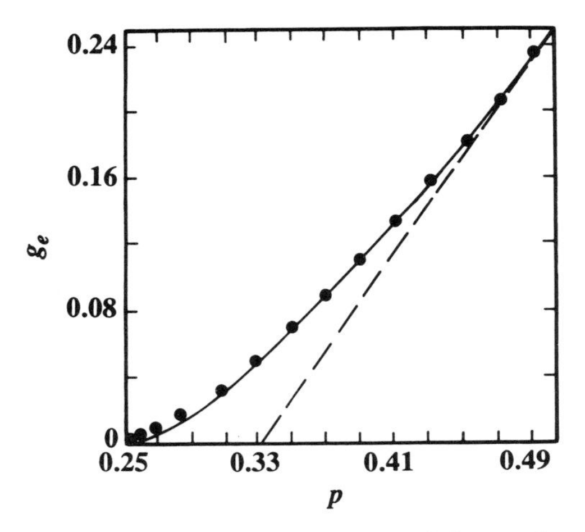

Figure 5.6 *Comparison of the REMA predictions of the conductivity g_e of a simple cubic network (curve) with the Monte Carlo data (circles). The dashed line represents the EMA predictions (after Sahimi et al. 1983b).*

5.5 Critical path method

Another method of estimating transport properties of disordered media was developed by Ambegaokar, Halperin, and Langer (AHL) in 1971, who argued that transport in a disordered medium with a *broad* distribution of conductances is dominated by those conductances whose magnitudes are larger than some characteristic value g_c, which is the largest conductance such that the set of conductances, $\{g_p \mid g_p > g_c\}$, forms a conducting sample-spanning cluster. Thus, all conductances that are smaller than g_c can be set to be zero, and transport in a disordered medium with a broad conductance distribution reduces to a percolation problem with threshold value g_c. These ideas were used for estimating hopping conductivity of semiconductors, discussed in Chapter 13.

Katz and Thompson (1986, 1987) extended the ideas of AHL to estimate the permeability and electrical conductivity of porous media. In a porous medium the local hydraulic conductance is a function of the pore length l, and therefore g_c defines a characteristic length l_c. Since both flow and electrical conduction problems belong to the class of scalar percolation problems (Chapter 2), the length that signals the percolation threshold in the flow problem also defines the threshold in the electrical conductivity problem. Thus, we have a trial solution for g_e given by

$$g_e = \phi g_c(l)[p(l) - p_c]^{\mu}, \tag{5.28}$$

where the porosity ϕ ensures a proper normalization of the fluid or the electric charge density. The function $g_c(l)$ is equal to $c_f l^3$ for the flow problem and $c_c l$ for the conduction problem. For appropriate choices of the function $p(l)$, the conductance $g_e(l)$ achieves a maximum for some $l_{max} \leqslant l_c$. In general l^f_{max} for the flow problem is different from l^c_{max} for the conduction problem, because the transport paths have different weights for the two problems.

If $p(l)$ allows for a maximum in the conductance, and if the maximum occurs for $l_{max} \leqslant l_c$, then we can write

$$l^f_{max} = l_c - \Delta l_f = l_c\{1 - \mu/[1 + \mu + l_c\mu p''(l_c)/p'(l_c)]\}, \tag{5.29}$$

$$l^c_{max} = l_c - \Delta l_c = l_c\{1 - \mu/[3 + \mu + l_c\mu p''(l_c)/p'(l_c)]\}. \tag{5.30}$$

If the pore size distribution of the medium is very broad, then $l_c\mu p''(l_c)/p'(l_c) << 1$, and using $\mu \simeq 2$ for three-dimensional systems yields

$$l^f_{max} = l_c[1 - \mu/[1 + \mu)] \simeq \frac{1}{3} l_c, \tag{5.31}$$

$$l^c_{max} = l_c[1 - \mu/[3 + \mu)] \simeq \frac{3}{5} l_c. \tag{5.32}$$

Now, writing

$$\sigma = a_1\phi[p(l^c_{max}) - p_c]^{\mu}, \tag{5.33}$$

and

$$k = a_2\phi(l^f_{max})^2[p(l^f_{max}) - p_c]^{\mu}, \tag{5.34}$$

we obtain to first order in Δl_c or in Δl_f,

$$p(l^{f,c}_{max}) - p_c = -\Delta l_{f,c}p'(l_c). \tag{5.35}$$

To obtain the constants a_1 and a_2, Katz and Thompson (1986) assumed that at a local level the rock conductivity is σ_f, the conductivity of the fluid that saturates the pore space, and that the local pore geometry is cylindrical. These imply that $a_2 = \sigma_f$ and $a_2 = 1/32$. Therefore, one obtains

$$k = a_3 l_c^2 \sigma/\sigma_f, \tag{5.36}$$

where $a_3 = 1/226$. A similar argument leads to

$$\frac{\sigma}{\sigma_f} = \frac{l^c_{max}}{l_c}\phi S(l^c_{max}), \tag{5.37}$$

where $S(l^c_{max})$ is the volume fraction (saturation) of connected pore space involving pore widths of size l^c_{max} and larger.

Equations (5.36) and (5.37) involve no adjustable parameters: every parameter is fixed and precisely defined. To obtain the characteristic length l_c,

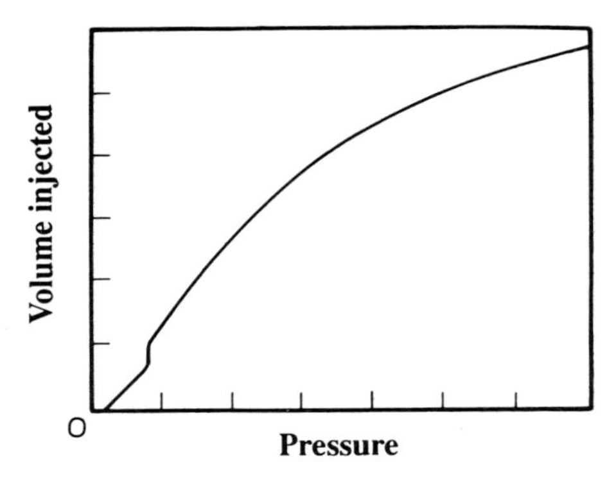

Figure 5.7 *A typical curve of the volume injected versus the pressure during mercury porosimetry. The knee denotes formation of a sample-spanning cluster.*

Katz and Thompson proposed mercury porosimetry discussed in Chapter 3. Consider a typical mercury porosimetry curve in which the pore volume of the injected mercury is obtained as a function of the pressure; see Fig. 5.7. The initial portion of the curve is obtained *before* a sample-spanning cluster of pores, filled with mercury, is formed. There is also an inflection point beyond which the pore volume increases rapidly with the pressure. This inflection point signals the formation of the sample-spanning cluster. Therefore, from the Washburn equation (3.2) we must have, $l \geqslant -4\sigma_i \cos\theta / P_i$, for the portion of the curve beyond the inflection point, where P_i is the pressure at the inflection point, and σ_i is the interfacial tension. Then, $l_c = -4\sigma_i \cos\theta / P_i$ defines the characteristic length l_c. Thus, the procedure to

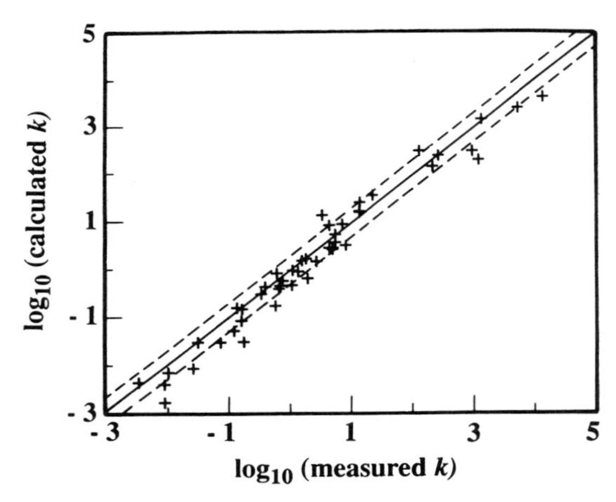

Figure 5.8 *Comparison of the percolation predictions of the permeability k (in millidarcy) of rocks with the data. Dashed lines denote a factor 2 of possible error (after Katz and Thompson 1986).*

use this method is as follows. From a mercury porosimetry experiment the parameter l_c is estimated, then (5.36) and (5.37) are used to estimate k and σ.

Figure 5.8 compares the permeability of a set of sandstones, predicted by (5.36), with the data. The dashed lines mark a factor of two. No adjustable parameter has been used and the agreement between predictions and theory appears to be very good. Similarly good agreement is obtained for the conductivity of the porous medium (Katz and Thompson 1987). Normally, l_c can be estimated from mercury injection curves with an error of at most 15%. The exponent μ used in (5.31) and (5.32) can take on its value for continuum percolation discussed in Chapter 2. This depends on the structure of the pore size distribution of the pore space.

The ideas of AHL and Katz and Thompson can also be extended for calculating the permeability of fractured rocks. Indeed, Charlaix *et al.* (1987) used arguments very similar to those of Katz and Thompson to calculate the permeability of fracture networks with a broad distribution of fracture apertures.

5.6 Random walk methods

Diffusion can be simulated by using random walk processes. A random walker, usually a massless particle, is put in the system to execute a random walk. If the porous medium is modeled by a random pore network with distributed pore conductances, then at each step of the walk the particle selects its next step with a probability proportional to the pore conductances. If a pore is closed (its conductance is zero), the walker is not allowed to move into that pore. We then repeat this random walk for a large number of particles, starting at randomly selected points of the system, and for a long enough time, where each time unit is equivalent to one step of the walk. The mean-square displacement $\langle r^2 \rangle$ of all walkers, in a d-dimensional network and after a long enough time t, is related to the effective diffusivity D_e by

$$\langle r^2 \rangle = 2dD_e t. \tag{5.38}$$

There are now three possibilities. (i) $\langle r^2 \rangle$ grows linearly with t. Then, (5.38) tells that D_e is a constant. This is the familiar ordinary diffusion known as Fick's law. (ii) $\langle r^2 \rangle$ grows with t *slower* than linearly. Thus, the motion of the particles is very slow and D_e *vanishes* at long times. We call this *sub-diffusion* or *anomalous* diffusion, or *fractal* diffusion, which is the situation that one encounters if $\langle r^2 \rangle^{1/2}$ is *less* than the correlation length or the dominant length scale of the system. Thus, diffusion is always very slow in fractal and self-similar media, since the dominant length scale of such systems is either infinite, or as large as their linear size. This regime of diffusion will be discussed in detail in Chapter 9. (iii) $\langle r^2 \rangle$ grows with t *faster* than linearly. Hence, the motion of the particles is very fast, and D_e *diverges* at long times. We call this *superdiffusion*, often encoutered in groundwater flow (see Chapter 6), turbulence, and many other practical situations. Subdiffusion and superdiffusion *cannot* be described by the classical continuum equation of diffusion with constant, or even time-dependent, diffusivity.

The first application of this idea for determining transport properties of disordered systems appeared in a paper of Haji-Sheikh and Sparrow (1966), who studied heat conduction in a composite solid. But the method was popularized by de Gennes (1976), who made an analogy between the motion of a random walker in a disordered medium and the motion of an ant in a labyrinth. Havlin and Ben-Avraham (1987) provide an extensive review of this subject. The computer algorithm for random walks in random conductance network models of porous media can be vectorized for use in computers such as Cray Y-MP, and it can be highly efficient; see Sahimi and Stauffer (1991) for details and a computer program.

If, instead of a random pore network, we have to deal with diffusion and flow in a disordered continuum of the type discussed in Chapter 2, then the problem is more complex. A traditional method, such as the finite-element technique, will be notoriously time-consuming for such systems, because even if we use only 20^3 grains (a modest number), a very fine finite-element mesh with roughly 10^9 nodes (and, thus, 10^9 equations) would be required to solve accurately a simple equation such as the Laplace equation for steady state conduction, a prospect which is totally impractical. For this reason alone, random walk methods are the preferred technique for estimating diffusivity and conductivity of porous media. One way of facilitating the random walk simulations in disordered continua is by using the *first passage time* (FPT) method. The idea is that if a random walker moves in a homogeneous region of the continuum, there is no need to spend unnecessary time to simulate detailed motion of the walker in that region. The walker can take long steps to quickly pass through the homogeneous region and arrive at the interface between the two phases. The necessary time for taking long steps can often be calculated *analytically*. In conventional simulations in random networks, each time a step is taken, the time is increased by one unit; in FPT, the walker takes long steps (as long as they do not take it outside of a phase) and the time is increased by an amount appropriate to that step. This basic idea was first used by Zheng and Chiew (1989) and Kim and Torquato (1990) for simulating reaction and diffusion in continuum models of porous media.

In an FPT simulation, we construct the largest (imaginary) concentric sphere of radius r, centered at the present position of the walker, which just touches the multiphase interface. The mean time t_m for the particle to reach a randomly selected point on the surface of the sphere can be calculated by using (5.38), $t_m(r) = r^2/(2d\sigma_i)$, where σ_i is the conductivity of the phase in which the particle is moving. This time is recorded, and the process of constructing the imaginary sphere and calculating the time that a point on its surface is reached by a random walker is repeated, until the random walker comes within a very small distance of the multiphase interface. We then compute the mean time necessary for crossing the boundary, t_b, and the probability of crossing the boundary, proportional to the ratio of the conductivities or diffusivities of the two phases. If the random walker crosses the interface and enters a new phase, it finds itself in a new homogeneous phase; the process of sphere construction is repeated and the mean-square displacement of the particle is computed which, after a long enough time, yields D_e and σ. This method is particularly effective for simulating transport in disordered continua. The efficiency of the method decreases as the porosity of the pore space decreases, since the search for the construction of the imaginary sphere becomes time-consuming, and near p_c the method is not efficient at all.

Another method of speeding up the random walk simulations is to use a *weak bias* in the simulations (Schwartz *et al.* 1989), such that the walker is more likely to move in the direction of macroscopic transport (or bias) than in the other directions. This bias causes the walker to sample the pore space more efficiently, because in the direction of the bias the traveled distance is proportional to N_s rather than $N_s^{1/2}$, where N_s is the number of steps.

Random walk methods are particularly useful for estimating the electrical conductivity of porous media comprised of an insulating granular matrix saturated with a conducting pore fluid. For example, Schwartz and Banavar (1989) used random walk simulations to calculate the electrical conductivity of the grain consolidation model of Roberts and Schwartz discussed in Chapter 3, with multisize granular particles. The results, shown in Fig. 5.9,

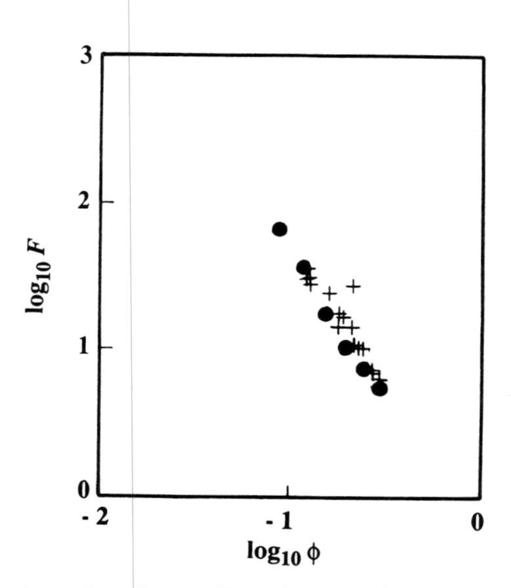

Figure 5.9 *Comparison of random walk predictions of the dependence of the formation factor F of a model consolidated porous medium on the porosity ϕ (+) with the experimental data (circles) (after Schwartz and Banavar 1989).*

in terms of the formation factor $F = \sigma_f/\sigma$, are in excellent agreement with the experimental data.

In contrast with D_e and σ, there is no random walk method for estimating k, because in general there is no exact relation between k and D_e, and therefore k cannot be related to the mean-square displacement of a random walker. Thus, estimating k has always been much more difficult than D_e or σ.

5.7 Conclusions

Effective-medium approximations are the simplest method of estimating transport properties of porous media. They are accurate over a wide range of porosity, and can easily provide an order of magnitude estimate of the properties of interest. But they are not accurate near p_c. Renormalization group methods can provide a more accurate description of the transport properties near p_c, but for three-dimensional systems the computations become complex. A simple combination of EMA and PSRG appears to provide the most accurate analytical approximation to the transport properties over the entire range of interest. The critical path methods are also very accurate for predicting permeability and conductivity of porous media, but they require as an input a parameter that can only be estimated by an experiment.

References

Ambegaokar, V., Halperin, B.I., and Langer, J.S., 1971, *Phys. Rev B.* **4**, 2612.
Benzoni, J. and Chang, H.-C., 1984, *Chem. Eng. Sci.* **39**, 161.
Bernasconi, J., 1974, *Phys. Rev. B* **9**, 4575.
Bernasconi, J., 1978, *Phys. Rev. B* **18**, 2185.
Bruggeman, D.A.G., 1935, *Annl. Phys.* **24**, 636.
Charlaix, E., Guyon, E. and Roux, S., 1987, *Trans. Porous Media*, **2**, 31.
de Gennes, P.G., 1976, *La Recherche* **7**, 919.
Doyen, P.M., 1988, *J. Geophys. Res.* **93**, 7729.
Haji-Sheikh, A. and Sparrow, E.M., 1966, *SIAM J. Appl. Math.* **14**, 370.
Harris, C.K., 1992, *Trans. Porous Media* **9**, 287.
Hasimoto, H., 1959, *J. Fluid Mech.* **5**, 317.
Havlin, S. and Ben-Avraham, D., 1987, *Adv. Phys.* **36**, 695.
Hinch, E.J., 1977, *J. Fluid Mech.* **83**, 695.
Katz, A.J. and Thompson, A.H., 1986, *Phys. Rev. B* **34**, 8179.
Katz, A.J. and Thompson, A.H., 1987, *J. Geophys. Res.* **B92**, 599.
Kim, I.C., and Torquato, S., 1990, *J. Appl. Phys.* **68**, 3892.
Kim, S. and Russel, W.B., 1985, *J. Fluid Mech.* **154**, 269.
Kirkpatrick, S., 1973, *Rev. Mod. Phys.* **45**, 574.
Koplik, J., Lin, C., and Vermette, M., 1984, *J. Appl. Phys.* **56**, 3127.
Landauer, R., 1952, *J. Appl. Phys.* **23**, 779.
Landauer, R., 1978, *AIP Conference Proceedings* **40**, 2.
Mendelson, K.S. and Cohen, M.H., 1982, *Geophysics* **47**, 257.

Odagaki, T. and Lax, M., 1981, *Phys. Rev. B* **24**, 5284.
Sahimi, M., 1993, *Rev. Mod. Phys.*
Sahimi, M., Hughes, B.D., Scriven, L.E., and Davis, H.T., 1983a, *J. Chem. Phys.* **78**, 6849.
Sahimi, M., Hughes, B.D., Scriven, L.E., and Davis, H.T., 1983b, *Phys. Rev. B* **28**, 307.
Sahimi, M. and Stauffer, D., 1991, *Chem. Eng. Sci.* **46**, 2225.
Sangani, A.S. and Acrivos, A., 1982, *Int. J. Multiphase Flow* **8**, 343.
Schwartz, L.M. and Banavar, J., 1989, *Phys. Rev. B* **39**, 11965.
Schwartz, L.M., Banavar, J., and Halperin, B.I., 1989, *Phys. Rev. B* **40**, 9155.
Sen, P.N., Scala, C., and Cohen, M.H., 1981, *Geophysics* **46**, 781.
Sheng, P., 1990, *Phys. Rev. B* **41**, 4507.
Stanley, H.E., Reynolds, P.J., Redner, S., and Family, F., 1982, in *Real-Space Renormalization*, edited by T.W. Burkhardt and J.M.J. van Leeuwen, Berlin, Springer, p. 169.
Torquato, S., 1991, *Appl. Mech. Rev.* **44**, 37.
Webman, I., 1981, *Phys. Rev. Lett.* **47**, 1496.
Zheng, L.H. and Chiew, Y.C., 1989, *J. Chem. Phys.* **90**, 322.

6
Hydrodynamic dispersion and groundwater flow in rock

6.0 Introduction

Chapter 5 discusses flow and transport processes in rock that involve only one fluid and one fluid phase. At the next level of complexity are those that involve at least two fluids but still one fluid phase. The most important of such phenomena is hydrodynamic dispersion, discussed in this chapter. Hydrodynamic dispersion is particularly important to groundwater flow pollution and for this reason has been studied for a long time.

6.1 The phenomenon of dispersion

When two miscible fluids are brought into contact, with an initially sharp front separating them, molecular diffusion gives rise to a transition zone across the initial front, the two fluids slowly diffuse into one another, and the original front becomes a diffused mixed zone whose composition changes from one pure fluid to the other. The net transport of one of the fluids across any arbitrary plane can be represented by the classical diffusion equation. This mixing process is independent of convective flows. However, if the fluids are flowing and the fluid velocity is not uniform, then some additional mixing of the two fluids is caused by the nonuniform velocity field, which in turn may be caused by the morphology of the medium, the fluid flow condition, and chemical or physical interactions with the solid surface of the medium. This mixing process, called *hydrodynamic dispersion*, is important to a wide variety of processes. Dispersion is important to miscible displacements used in enhanced recovery of oil. When oil production from a reservoir declines, a fluid is injected into the reservoir to mobilize the oil

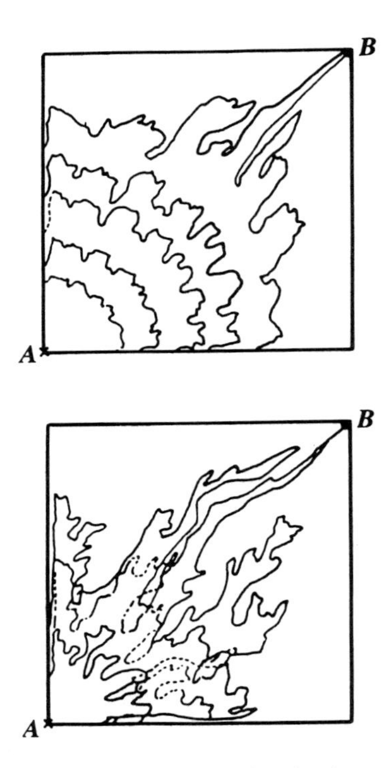

Figure 6.1 *Miscible displacement patterns. The displacing fluid is injected at A to produce the displaced fluid at B. Different fronts are shown with increasing time. The top diagram is for $M = 2.5$, where M is the ratio of the viscosities of the displaced and displacing fluids; the bottom diagram is for $M = 17.5$.*

and increase the production. The injected fluid can be immiscible or miscible with the oil in place. Injection of immiscible fluid produces a two-phase flow problem, discussed in the next chapter. If the viscosity of the displacing miscible fluid is less than that of oil, then large fingers of the displacing fluid are formed that advance thoughout the medium, and leave behind a large amount of oil; see Fig. 6.1. Dispersive mixing of the displacing fluid and the oil can help the fingers join and displace more oil, thus increasing the efficiency of the process. Dispersion is also important to saltwater intrusion in a coastal aquifer, where fresh and salt waters mix by a dispersion process, to in situ study of the characteristics of an aquifer, where a classical method of determining such characteristics is injecting fluid tracers in it to mix with the water and measuring their travel times between two points, to the pollution of the subsurface water because of industrial and nuclear wastes, and to flow and reaction in packed-bed chemical reactors.

Dispersion in flow through a porous medium is characterized by the *first passage time distribution* (FPTD) of tracer particles traveling in the flowing fluid. Suppose that we inject into a porous medium a population of fluid particles that are miscible with the flowing fluid in the medium. This population of particles passes the entrance plane to the medium at the same instant, but will mix with the flowing fluid and arrive at the exit plane, *for the first time*, with a distribution of transit times, called the FPTD, because each fluid particle selects a different path for its journey, and therefore spends a different amount of time to arrive at the exit plane. If the FPTD is measured at the exit plane and in the direction of the macroscopic flow, then its broadness is a measure of *longitudinal dispersion* in the medium, i.e., mixing of the two fluids and spreading of the injected fluid in the direction of the macroscopic flow. However, a population of particles entering the medium at the same point will not follow entirely the mean flow to the exit plane, but will also be dispersed in the transverse directions (perpendicular to the macroscopic flow) with a wider distribution of exit locations than that of entrance locations. The associated FPTD, measured at the same transverse plane is a measure of *transverse dispersion* in a porous medium. If the macroscopic flow is unidirectional, then, the mean flow velocity in the transverse direction is zero, and therefore transverse dispersion (mixing) is *always* smaller than longitudinal dispersion.

Two basic mechanisms drive dispersion in macroscopically homogeneous, microscopically disordered porous media. The first mechanism is *kinematic*: streamlines along which the fluid particles travel divide and rejoin repeatedly at the junctions of flow passages in the pore space. The consequent tangling and divergence of streamlines is accentuated by the widely varying orientations of flow passages and the coordination numbers of the pore space. The result is a wide variation in the lengths of streamlines and the downstream transverse separations of the streamlines. The second mechanism is *dynamic*: the speed with which a given flow passage is traversed depends on the flow resistance or hydraulic conductance of the passage, its orientation, and the local pressure field. The two mechanisms conspire to produce broad FPTDs, and hence extensive convective mixing.

These two fundamental mechanisms of dispersion do *not* depend on molecular diffusion. Diffusion modifies convective mixing not only by transferring the particles into and out of the stagnant regions of the pore space, but also by moving them out of the slow boundary layers that are usually formed near the solid surfaces. Thus, the modification of dispersion by diffusion depends on pore space morphology and how it in turn affects local flow and concentration fields. This discussion should give the reader some feeling about the significance and complexity of dispersion processes. Among the various flow phenomena occuring in porous media, dispersion is perhaps the one most capable of application in different fields, as is evident from the foregoing practical applications.

6.2 The convective diffusion equation

One of the most important tasks in modeling dispersion is developing the appropriate equation for the evolution of the solute (fluid particles) concentration. Traditional modeling of dispersion processes in microscopically disordered and macroscopically homogeneous porous media is usually based on the convective diffusion equation (CDE)

$$\frac{\partial C}{\partial t} + \boldsymbol{v} \cdot \nabla C = D_L \frac{\partial^2 C}{\partial x^2} + D_T \nabla_T^2 C, \tag{6.1}$$

where $\boldsymbol{v}$ is the macroscopic mean velocity, C the macroscopic mean concentration of the solute, and ∇_T^2 the Laplacian in transverse directions. Thus, dispersion is modeled as anisotropic diffusion of the solute or the injected fluid augmented by the flow field, represented by $\boldsymbol{v} \cdot \nabla C$; the diffusivities are the longitudinal dispersion coefficient D_L (in the direction of mean flow) and the transverse dispersion coefficient D_T. The anisotropy is dynamic and is caused by the flow field since the fluid particles mix with and spread in the flowing fluid better in the longitudinal direction than the transverse direction. Dispersion is said to be *macroscopically diffusive* or Gaussian if it obeys the CDE. If a particle population is injected into the medium at the point $\mathbf{r}_0 = (x_0, y_0, z_0)$ at time $t = 0$, for macroscopically diffusive dispersion the probability density $P(\mathbf{r},t)$ obeys the Gaussian distribution

$$P(\mathbf{r}, t) = (8\pi^3 D_L D_T^2 t)^{-3/2} \exp\left[-\frac{(x - x_0 - v\,t)^2}{4D_L t} - \frac{(y - y_0)^2}{4D_T t} - \frac{(z - z_0)^2}{4D_T t}\right], \tag{6.2}$$

where $P(\mathbf{r}, t)d\mathbf{r}$ is the probability that a particle is in a plane between $\mathbf{r}$ and $\mathbf{r} + d\mathbf{r}$ at time t, $v = |\boldsymbol{v}|$, and $\mathbf{r} = (x, y, z)$. $P(\mathbf{r}, t)$ is proportional to C/C_0, where C_0 is the concentration at $t = 0$. If we define $Q(\zeta - \zeta_0, t)dt$ as the probability that a particle, beginning in the plane at ζ_0 will cross, *for the first time*, a plane at ζ between t and $t + dt$, then from (6.2) we can easily obtain the FPTD

$$Q(\zeta - \zeta_0, t) = |\zeta - \zeta_0|(4\pi D_\zeta t^3)^{-1/2} \exp[-(\zeta - \zeta_0 - v_\zeta t)^2/4D_\zeta t], \tag{6.3}$$

where D_ζ and v_ζ are the dispersion coefficient and the mean flow velocity in the ζ-direction, respectively. Various moments of Q yield information about the flow field and the dispersion processes. For example, for the longitudinal direction we have

$$\langle t \rangle = \frac{L}{v}, \tag{6.4}$$

and

$$\langle t^2 \rangle = \langle t \rangle^2 \left(1 + \frac{2D_L}{Lv}\right), \tag{6.5}$$

where $L = \zeta - \zeta_0$. In general, we can easily show that, for diffusive dispersion and large L, and to the leading order we have, $\langle t^n \rangle \sim \langle t \rangle^n$ where $n > 1$ is any integer number. Therefore, one way of showing that the CDE cannot describe a dispersion process is to show that $\langle t^n \rangle / \langle t \rangle^n$ $(n > 1)$ depends on t and is *not* a constant.

The description of dispersion in terms of the CDE is purely phenomenological. It provides no insight into how D_L and D_T depend on the morphology of the pore space. For this reason, there have been many studies of dispersion processes in rocks using a wide variety of techniques and models of pore space that vary anywhere from a single capillary tube to a complex network of pores and/or fractures. We shall only discuss the most important results to which percolation is relevant. The interested reader is referred to Sahimi (1993) for a more complete discussion of this subject. We first discuss the most important experimental facts about dispersion in porous media, so that we have a clear picture of what we are supposed to model and predict.

6.3 Classification of dispersion in porous media

Experimental studies have shown (for a review see Sahimi 1993) that dispersion processes in consolidated and unconsolidated porous media are similar, and thus we do not need to distinguish between them. If we define a Péclet number, $Pe = d_g v / D_m$, where d_g is the average diameter of a grain, and D_m is the molecular diffusivity, then experimental data show that, depending on the value of Pe, we may have several distinct regimes of dispersion.

(i) $Pe < 5$ is the pure diffusion regime, because convection is so slow that diffusion controls dispersion almost completely. Moreover, it can be shown that

$$\frac{D_L}{D_m} = \frac{D_T}{D_m} = \frac{1}{F\phi}, \tag{6.6}$$

where, as in Chapter 5, F is the formation factor of rock and ϕ is its porosity. The quantity $1/(F\phi)$ varies commonly between 0.15 and 0.7, depending on the porous medium.

(ii) $5 < Pe < 300$ is the regime where convection dominates dispersion, but the effect of diffusion cannot be neglected; we have

$$\frac{D_L}{D_m} \sim Pe^{\beta_L}, \tag{6.7}$$

$$\frac{D_T}{D_m} \sim Pe^{\beta_T}. \tag{6.8}$$

The *average* values of β_L and β_T from all the available experimental data are, $\beta_L \simeq 1.2$ and $\beta_L \simeq 0.95$. We call this regime the *boundary layer* dispersion (Koch and Brady 1985), since in this regime the main role of diffusion is to transfer the solute out of the slow boundary layer zones near the solid surface. This regime of dispersion was first discussed by Saffman (1959), who also showed that

$$\frac{D_L}{D_m} \sim Pe \ln Pe, \tag{6.9}$$

$$\frac{D_T}{D_m} \sim Pe. \tag{6.10}$$

It can be shown that (6.7) and (6.9) are consistent with each other, because $Pe \ln Pe$ can be accurately represented by Pe^{β_L}.

(iii) $300 < Pe < 10^5$ is the regime where the effect of diffusion is negligible and one has pure convective mixing. Simple dimensional analysis indicates that,

$$\frac{D_L}{D_m} \sim Pe, \tag{6.11}$$

$$\frac{D_T}{D_m} \sim Pe. \tag{6.12}$$

This is usually called *mechanical dispersion*. In this case dispersion is simply the result of a stochastic velocity field induced by the randomly distributed effective pore radii. The same is true about dispersion in macroscopically heterogeneous rock where the stochastic velocity field is caused by the spatial variations of the permeability.

(iv) Finally, there is a distinct dispersion regime, the so-called *holdup dispersion* (Koch and Brady 1985), first studied by Turner (1959) and Aris (1959). In this regime the solute is trapped in a dead-end or stagnant region, or inside the solid grains, from which it can escape only by molecular diffusion. We have

$$\frac{D_L}{D_m} \sim Pe^2, \tag{6.13}$$

$$\frac{D_T}{D_m} \sim Pe^2. \tag{6.14}$$

In this regime, the strength of mixing due to diffusion into and out of the stagnant zones is comparable to the convective mixing. In a porous medium near its percolation threshold p_c, there are many such dead-end pores or stagnant zones, and therefore this regime is relevant to such a porous medium. The above results have been obtained and confirmed by numerous authors. For example, Bacri *et al.* (1987) measured D_L for three different porous media and showed that disorder strongly affects D_L and its dependence on *Pe*. They were able to observe power laws (6.7)–(6.14), depending on the broadness of the pore size distribution and the connectivity of the pore space.

Let us discuss three experimental studies that directly relate dispersion to percolation. Charlaix *et al.* (1988) carried out dispersion experiments in two-dimensional hexagonal networks of pores whose effective diameters were of the order of millimeters. They found that as the fraction of open pores decreased, D_L increased sharply, and that (6.7) was obeyed. But even when D_L was measured quite close to the p_c of the medium, (6.13) was *not* obeyed, presumably because the exchange time between the flowing fluid and the dead-end pores was so large that its effect could not be detected during the experiment. Hulin *et al.* (1988) measured D_L in bidispered sintered glass materials prepared from mixtures of two sizes of beads. When the porosity was decreased from 30% to 12%, they observed that D_L increased by a factor of 30. These two studies also indicated that dispersion is more sensitive to *large-scale* inhomogeneities than to the *local* structure of a porous medium. Finally, Gist *et al.* (1990) measured D_L in a variety of natural and synthetic porous media and observed that it sharply increases as the porosity of the medium decreases. We now discuss two percolation models that can explain all of these data.

6.4 Network and percolation models of dispersion

Sahimi *et al.* (1983) were the first who used percolation and random network models of porous media (Chapter 5) for simulating dispersion. Their method first determines the flow field in the network. A macroscopic pressure gradient is applied to the network in one direction, and periodic or cyclic conditions are used in the other directions. The total mass of the fluid reaching any node of the network is a conserved quantity. Thus, assuming that the pores are cylindrical tubes whose radii are selected from a distribution function, and that one has laminar flow in the pores, we can write, $q_{ij} = \pi \Delta P R_{ij}^4/(8\eta l)$, for the volumetric flow rate q_{ij} of pore ij with radius R_{ij} and length l, along which a pressure drop ΔP has been imposed, where η is the fluid viscosity. Thus, for each node i we can write

$$\sum_{ij} q_{ij} = 0, \tag{6.15}$$

where the sum is over all pores ij connected to node i. Writing this equation for all internal nodes of the network results in a set of N simultaneous linear equations for the nodal pressures of a network of N internal nodes, which can be solved numerically by a direct or an iterative method. From the solution of the set, the pore flow rates and mean velocities $v_m = \Delta P R_{ij}^2/(8\eta l)$, are calculated.

The CDE implies that dispersion can be thought of as anisotropic diffusional mixing of a solute with a miscible solvent augmented by a flow field. Since diffusion can be simulated by a random walk process (Chapter 5), we can also model dispersion by a random walk. Thus, after calculating the flow field throughout the network, a large number of fluid particles are injected into the network at random positions at the upstream plane, $x = 0$. Each particle selects a streamline at a radial position r, within a pore of radius R, whose speed is determined from, $v_p = 2v_m(1 - r^2/R^2)$, and travels along this streamline. The convective travel time for a given pore is given by, $t = t_c = l/v_p$. Complete mixing at the nodes of the network is assumed, and therefore, once a fluid particle passes through the pore and arrives at a node, it completely forgets its past history. Then, a new pore is selected with a probability that, because of complete mixing at the node, is proportional to the flow rate in that pore. Thus, pores that are nearly in the direction of macroscopic flow have the highest probability of being selected, but those pores that are nearly perpendicular to the direction of macroscopic flow can also be selected, albeit with a much smaller likelihood. No nodal residence time is assumed (i.e., the fluid particle leaves the nodes instantly). A streamline in the new pore is selected, the travel time along which is computed, and so on. The total travel time of a fluid particle is simply the sum of its travel times in various pores. The FPTDs for the particles are computed by fixing the longitudinal and transverse positions and measuring the time t at which the particles arrive at these positions for the first time, from which D_ζ, the dispersion coefficient in the ζ-direction is calculated

$$D_\zeta = \left\langle \frac{\sigma_\zeta^2}{2t} \right\rangle, \tag{6.16}$$

where $\zeta_0(x_0)$ is the starting position of the particles, $\sigma_x^2 = (x - x_0 - vt)^2$, $\sigma_\zeta^2 = (\zeta - \zeta_0)^2$ for $\zeta = y$ or z, and the averaging is taken over all particles. Equation (6.16) is written in analogy with random walk processes discussed in Chapter 5. Using (6.3), it can also be shown that $D_L = v^3(\langle t^2 \rangle - \langle t \rangle^2)/(2L)$ for a network of linear size L.

In this model both D_L and D_T depend linearly on v, because if we double the macroscopic pressure gradient, we also double the flow velocities of the pores, and thus dispersion coefficients are also doubled. This is in agreement with (6.11) and (6.12), therefore this model is appropriate for mechanical dispersion, since pore-level molecular diffusion has been ignored. Pore-level molecular diffusion is important to dispersion in the boundary layers near the pore surface. To include the effect of molecular diffusion and simulate boundary layer dispersion, the following method is adopted (Sahimi and Imdakm 1988). The convective time t_c for travelling along a streamline is first calculated. If the streamline along which the particle intends to travel is close to the pore walls, then $t_c \gg t_r$, where t_r is the radial

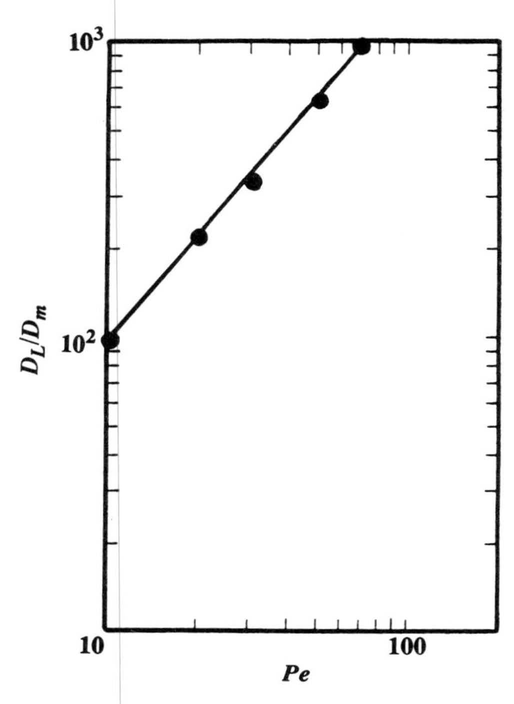

Figure 6.2 *The Péclet number Pe dependence of the longitudinal dispersion coefficient D_L, as predicted by the percolation model. D_m is the molecular diffusivity. The slope of the line is about 1.2 (after Sahimi and Imdakm 1988).*

diffusion time scale, of the order of $R^2/(8D_m)$, the time for diffusion to a streamline at a radial position $r = R/2$. Then, one sets, $t = t_c + t_r$, since the fluid particle has enough time to diffuse to a faster streamline. Otherwise, diffusion in the pore is ignored. As soon as such diffusion time scales are included in the total travel time of the particle, the linear dependence of D_L and D_T on v is destroyed, because diffusion time scales are independent of the flow field. To simulate holdup dispersion, the fluid particles are allowed to diffuse into the dead-end pores of the network. Mixing in dead-end pores is only by molecular diffusion, therefore the time to travel along such pores is $t_{de} = l^2/(2D_m)$. In a series of papers, Sahimi *et al.* (1983, 1986) and Sahimi and Imdakm (1988) showed that these network models can accurately predict the most significant features of dispersion, in particular (6.9)–(6.14). For example, Fig 6.2 presents the dependence of D_L/D_m on Pe. The slope of the line is about 1.2, which agrees well with the data discussed above.

Koplik *et al.* (1988) proposed another method in which a one-dimensional CDE is assumed to hold for each pore

$$\frac{\partial C}{\partial t} + v_m \frac{\partial C}{\partial x} = D_m \frac{\partial^2 C}{\partial x^2}. \qquad (6.17)$$

Consider now a network of pores $\{ij\}$. The concentration C_{ij} in each pore obeys (6.17). At each internal node of the network (not those at which the external pressures are specified), we have conservation of mass, which means that

$$\sum_{\{ij\}} S_{ij}\left(v_{ij}C_{ij} - D_m \frac{\partial C_{ij}}{\partial x_{ij}}\right)_{x_{ij}=0} = 0, \tag{6.18}$$

where S_{ij} is the cross-sectional area of pore ij, and v_{ij} the mean flow velocity in that pore. Equation (6.17) is easily solved in Laplace transform space. Denoting the Laplace transform of $C_{ij}(x, t)$ by $\hat{C}_{ij}(x, \lambda)$, the solution is given by

$$\hat{C}_{ij}(x, \lambda) = A_{ij}\exp(\alpha_{ij}x) + B_{ij}\exp(\beta_{ij}x), \tag{6.19}$$

$$\alpha_{ij}, \beta_{ij} = [v_{ij} \pm (v_{ij}^2 + 4D_m\lambda)^{1/2}]/(2D_m), \tag{6.20}$$

where the coefficients A_{ij} and B_{ij} are determined from the boundary conditions for the pores, which are $C_{ij}(0, t) = C_{0ij}(t)$, and $C_{ij}(l, t) = C_{lij}(t)$. Writing down (6.18) for every internal node of the network yields a set of simultaneous linear equations for the (Laplace transformed) nodal concentrations (remember that C_{0ij} and C_{lij} are unknown). The set of linear equations for nodal concentrations is solved for several values of the Laplace transform variable λ, and the inverse Laplace transform of the numerical solution for the concentration field is obtained numerically. From this solution the dispersion coefficients can be calculated. This model too can predict certain features of experimental data on dispersion discussed above. Roux *et al.* (1986) proposed the same model but solved the governing equations by a different method.

6.5 Comparison with experimental data

The model of Sahimi *et al.* can predict the most important features of dispersion in porous media. Moreover, the network and percolation models described above predict that as p_c is approached, D_L and D_T increase sharply. If we define the *dispersivities* by $\alpha_L = D_L/v$ and $\alpha_T = D_T/v$, then an increase in D_L or D_T implies an even larger increase in the dispersivities, because as p_c is approached the average fluid velocity decreases ($v = 0$ at p_c). Figure 6.3 shows the increase of D_L in flow through a percolating square network as p_c is approached. This dramatic increase of D_L is due to the fact that near p_c the paths of the fluid particles are very tortuous, since many pores are closed to flow; this results in a broad FPTD and large D_L. This is in complete agreement with the experimental observations of Charlaix *et al.* (1988), Hulin *et al.* (1988), and Gist *et al.* (1990). The increase in α_L can

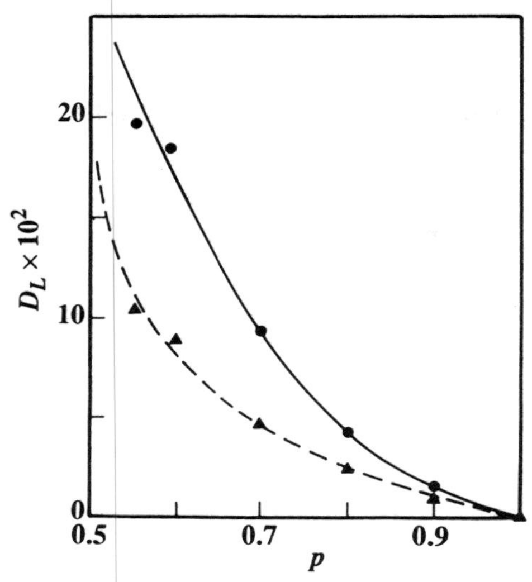

Figure 6.3 *Variations of the longitudinal dispersion coefficient D_L with the fraction of open bonds p in a square network. Circles denote the results for the case in which diffusion into the dead-end pores has been taken into account; triangles show the results for dispersion in the backbone only (after Sahimi and Imdakm 1988).*

also be understood by the following argument. Physically, α_L represents the macroscopic length scale for the homogeneity of the system and the applicability of the CDE for describing dispersion. As p_c is approached, the correlation length ξ_p of percolation increases. For length scales $L << \xi_p$ the sample-spanning cluster and its backbone are fractal objects, in which case a continuum equation such as the CDE cannot describe dispersion. Therefore, ξ_p is similar to α_L, and an increase in ξ_p should also cause a corresponding increase in α_L.

Gist *et al.* (1990) used percolation concepts and the model of Sahimi *et al.* to quantitatively predict their own data. Using (2.4) and (2.6), we can write, $\xi_p/d_g \sim (X^A)^{-\nu_p/\beta_p}$. Since for three-dimensional percolation networks, $\nu_p/\beta_p \simeq 2$, and because X^A is roughly proportional to the saturation (volume fraction) of the fluid S, we obtain $\xi_p/d_g \sim S^{-2}$. If we now define $\alpha_r = \alpha_L/d_g$, then the numerical simulations of Gist *et al.* (1990), using a method similar to that of Sahimi *et al.*, incidate that

$$\alpha_r \sim \left(\frac{\xi_p}{d_g}\right)^{2.2} \sim S^{-4.4}. \tag{6.21}$$

Although their data, shown in Fig. 6.4, have considerable scatter, the power law exponent obtained from the data agrees well with (6.21). Thus, percolation models can provide quantitative predictions for the experimental data on dispersion.

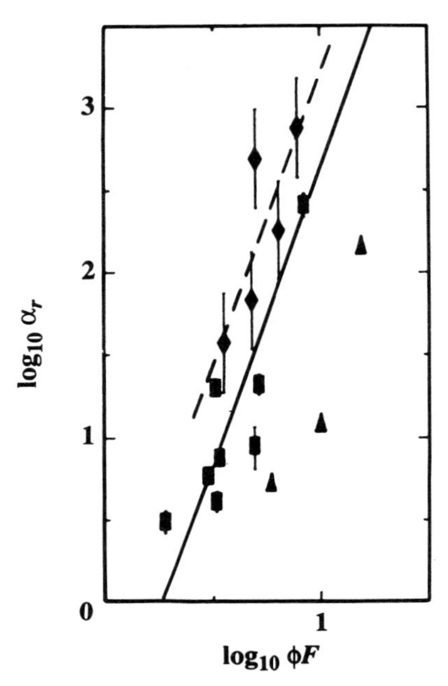

Figure 6.4 *Variations of α_r, the ratio of longitudinal dispersivity and the grain size, with the product of the porosity ϕ and the formation factor F, for sandstones (squares), epoxies (triangles), and carbonates (diamonds). The solid line is the fit of the data for sandstones, while the dashed line is that of the carbonates (after Gist et al. 1990).*

6.6 Percolation models of scale-dependent dispersion in heterogeneous rock

We now discuss dispersion in heterogeneous porous media, those with large-scale, spatially varying properties such as permeability. This problem has attracted considerable attention from hydrologists and *politicians*, as a result of growing concerns about pollution and water quality. Intensifying exploitation of groundwater and the increase in solute concentrations of aquifers, due to saltwater intrusion, leaking repositories, and use of fertilizers, have made dispersion in heterogeneous porous media a main topic of research. Dispersion is the main mechanism of carrying the pollutants and mixing them with unpolluted water.

The above network models of dispersion can also be used for macroscopically heterogeneous media. In this case, each bond of the network represents a region of the pore space over which the medium is homogeneous. Thus, the permeability that we assign to a bond is that of the region and not of a pore. In macroscopically heterogeneous rock, the permeabilities of various regions of the rock are not distributed randomly but are usually highly correlated. The permeability of the bonds should be assigned in a

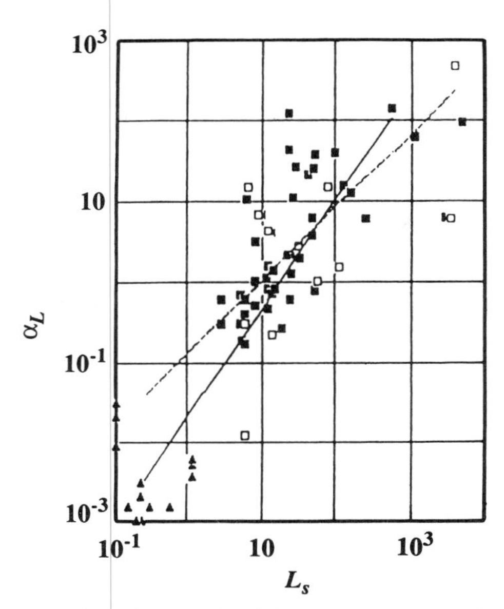

Figure 6.5 *Dependence of the longitudinal dispersivity α_L on the distance L_s from the injection point. Both quantities are in meters (after Arya et al. 1988).*

correlated manner, in which the range and type of correlations should be compatible with the available experimental data.

Pore-level diffusion does not play an important role in dispersion in macroscopically heterogeneous media, since dispersion is caused principally by large-scale fluctuations in the velocity field, and the fluctuations are caused by the spatial variations of the permeability field. Dispersion in fractured rocks can also be modeled by methods very similar to those we have discussed here. This is because large-scale fracture networks are similar to percolation networks (Chapter 4).

There have been several field studies of dispersion, e.g., those by Molz *et al.* (1983) and Sudicky *et al.* (1985), indicating that the dispersion coefficients and dispersivities are scale-dependent. Arya *et al.* (1988) collected over 130 dispersivities α_L from the literature. They vary anywhere from less than 1 mm to over 1 km, collected on length scales that vary from less than 10 cm to more than 100 km; see Fig. 6.5. The data show large scatter, but the straight line obtained by regression analysis and drawn through the data indicates that at least 75% of the data do follow the regression line with 95% confidence. The data shown in Fig. 6.5 indicate that

$$\alpha_L \sim L_s^{0.75}, \tag{6.22}$$

where L_s is the length scale of measurements or distance from the source (where tracer particles are injected into the porous medium). Neuman

(1990) presented another regression analysis of these and other data. He proposed that there are in fact two regimes, one for $L_s << 100$ m, for which the exponent in (6.22) was found to be about 1.5, and another one for $L_s >> 100$ m for which the exponent was found to be about 0.75, in agreement with (6.22). Before the application of percolation, and especially fractal, concepts to porous media became popular, (6.22) was considered a mystery, because classical models based on a CDE could not predict scale-dependent dispersivities. We show that percolation and fractals provide a simple explanation for (6.22).

To explain these data, a fractal approach was developed by Hewett (1986), Philip (1986), Arya *et al.* (1988), Ababou and Gelhar (1990), and Neuman (1990). Consider a stationary stochastic process $B_H(x)$ with the following mean and variance

$$\langle B_H(x) - B_H(x_0)\rangle = 0, \tag{6.23}$$

$$\langle [B_H(x) - B_H(x_0)]^2\rangle \sim |x - x_0|^{2H}, \tag{6.24}$$

where x and x_0 are two arbitrary points, and H is called the Hurst exponent (Mandelbrot and Van Ness 1968). This stochastic process is called *fractional Brownian motion* (fBm). The usual Brownian motion or random walk corresponds to $H = 1/2$. A remarkable property of fBm is that it generates correlations that are essentially of infinite extent. For example, if a correlation function is defined by

$$C(x) = \frac{\langle -B_H(-x)B_H(x)\rangle}{\langle B_H(x)^2\rangle}, \tag{6.25}$$

we find that $C(x) = 2^{2H-1} - 1$, i.e., this correlation function is *independent* of x. Moreover, the type of correlations can be tuned by varying H. If $H > 1/2$, then fBm displays *persistence*, i.e., a trend (a high or low value of the variable) at x is likely to be followed by a similar trend at $x + \Delta x$, whereas if $H < 1/2$, then fBm generates *antipersistence*, i.e., a trend at x is not likely to be followed by a similar trend at $x + \Delta x$. For $H = 1/2$ there are no correlations of the above type, and the overall shape of fBm traces is similar to that of a random walk. Thus, varying H allows us to generate highly correlated or anticorrelated heterogeneities by fBm. The above one-dimensional distribution can be easily extended to higher dimensions. We write

$$\langle [B_H(\mathbf{r}) - B_H(\mathbf{r}_0)]^2\rangle \sim |\mathbf{r} - \mathbf{r}_0|^{2H}, \tag{6.26}$$

where $\mathbf{r} = (x, y, z)$ and $\mathbf{r}_0 = (x, y, z)$. A convenient method of representing a distribution function is through its spectral density $S(\mathbf{k})$ which is the Fourier transform of its variance. For fBm in d dimensions it can be shown that

$$S(\mathbf{k}) \sim \frac{1}{\left(\sum_i^d k_i^2\right)^{H+d/2}}, \tag{6.27}$$

where $\mathbf{k} = (k_1 \ldots k_d)$. Note that fBm is not differentiable, but it can be made so by smoothing it over an interval. Using this technique, the derivative of fBm, called *fractional Gaussian noise* (fGn), can be obtained whose spectral density in, e.g., one dimension, is given by

$$S(k) \sim \frac{1}{k^{2H-1}}. \tag{6.28}$$

Hewett (1986) analyzed vertical porosity logs of many reservoirs and found that they follow fGn with $H \simeq 0.7$–0.8, indicating long range positive correlations. Moreover, the permeabilities follow fBm with essentially the same value of H as that found for the porosities. Hewett (1986) and others then analyzed dispersion in heterogeneous porous media with such distributions of permeabilities and porosities, and argued that the dispersivity α_L increases with time as

$$\alpha_L \sim t^{2H-1}, \tag{6.29}$$

so that with $H \simeq 0.75$ we obtain $\alpha_L \sim t^{0.5}$. This equation appears to provide accurate predictions for the available data. We now show that a percolation model of dispersion can predict essentially the same equation.

Chapter 2 discusses the scaling behavior of the permeability k and diffusivity D_e of a percolating system near p_c. What are the scaling laws for D_L near p_c? We define a time scale τ_s such that for times $t \gg \tau_s$ dispersion is diffusive and follows a CDE, whereas for $t \ll \tau_s$ dispersion is not Gaussian, with a crossover taking place at about $t \simeq \tau_s$. For dispersion near p_c, this time scale can be estimated from

$$\tau_s \sim \frac{\xi_p^2}{D_L}, \tag{6.30}$$

and therefore we must obtain the scaling of D_L near p_c in order to obtain the scaling laws for τ_s. But since mechanical dispersion seems to be the dominant mechanism in heterogeneous rocks, it is sufficient to study this limit. Assuming that the macroscopic flow is in the x direction, we introduce a *random walk fractal dimension* d_w defined by

$$\langle \Delta x^2 \rangle \sim t^{2/d_w}, \tag{6.31}$$

where $\langle \Delta x^2 \rangle = \langle (x - \langle x \rangle)^2 \rangle = \langle x^2 \rangle - \langle x \rangle^2$. If dispersion is diffusive, then $d_w = 2$, and the description of dispersion by a CDE is adequate. But we may also have $d_w > 2$, in which case $\langle \Delta x^2 \rangle$ grows with time slower than linearly (subdiffusion), and therefore D_L is *zero* after a long enough time (6.16). Alternatively, we can have $d_w < 2$, in which case $\langle \Delta x^2 \rangle$ grows with time faster than linearly (superdiffusion), and D_L will *diverge* after a long enough time. In both of these cases dispersion is not described by a CDE.

For dispersion in a percolating network two average flow velocities can be considered. If the fluid particles spend a considerable amount of time in the dead-end pores (holdup dispersion), then the average velocity should be defined in terms of the total travel time in the network (dead-end pores plus the backbone). Then, near p_c the average particle velocity v_c in the network scales as, $v_c \sim k/X^A$ (remember that only a fraction X^A of the pores are in the sample-spanning cluster). Using (2.4) and (2.11) we obtain

$$v_c \sim (p - p_c)^{\mu - \beta_p} \sim \xi_p^{-\theta}, \tag{6.32}$$

where, $\theta = (\mu - \beta_p)/\nu_p$. On the other hand, if the fluid particles spend a negligible amount of time in the stagnant or dead-end pores (mechanical dispersion), then the average particle velocity should be defined only in terms of the travel times in the backbone. The average particle velocity v_B in the backbone is given by $v_B \sim k/X^B$, or

$$v_B \sim (p - p_c)^{\mu - \beta_B} \sim \xi_p^{-\theta_B} \tag{6.33}$$

where, $\theta_B = (\mu - \beta_\beta)/\nu_p$. We also define a *macroscopic* Péclet number

$$Pe_m = \frac{v\xi_p}{D_e}, \tag{6.34}$$

where, depending on whether we have holdup or mechanical dispersion, v can be either v_c or v_B. Chapter 2 shows that, for $L \ll \xi_p$, we replace ξ_p in the above equations by L. Having defined these, we can now investigate the scaling of D_L and τ_s near p_c.

If the permeabilities of a heterogeneous rock are distributed according to fBm, then their distribution is very broad. Consequently, there should be regions of very low permeabilities that contribute very little to fluid flow in the rock. Mechanical dispersion seems to be operative in a heterogeneous rock, so the low permeability zones should not contribute to dispersion. Following the critical path method discussed in Chapter 5, we eliminate such low permeability zones from the system, and thus obtain a percolating backbone. According to (6.11) for mechanical dispersion we have $D_L/D_m \sim Pe$. It is not unreasonable to replace D_m, the pore-level molecular diffusivity, with D_e, the network-level effective diffusivity, and Pe, the local Péclet number, with Pe_m, the macroscopic one. Therefore, we obtain $D_L \sim \xi_p v_B \sim \xi_p^{1-\theta_B} \sim (p - p_c)^{\mu - \beta_B - \nu_p}$. Using the numerical values of μ, β_B and ν_p given in Table 2.3, we obtain, $D_L \sim (p - p_c)^{-0.56}$ in two dimensions and $D_L \sim (p - p_c)^{0.04}$ in three dimensions. This demonstrates the strong effect of the backbone structure on dispersion. Chapter 2 shows the backbone can be approximated by links and blobs. The blobs are very dense in two dimensions, providing a wide variety of paths for the fluid particles with a broad FPTD. As a result, D_L diverges as p_c is approached. On the other hand, the blobs are not very dense in three dimensions, which means that the FPTD is not broad enough to give rise to a divergent D_L. For $L \ll \xi_p$, we replace ξ_p with L to obtain, $D_L \sim L^{1-\theta_B}$. On the other hand, $D_L \sim d\langle \Delta x^2 \rangle / dt \sim L^{1-\theta_B} \sim \langle \Delta x^2 \rangle^{(1-\theta_B)/2}$. A simple integration then yields $\langle \Delta x^2 \rangle \sim t^{2/(1+\theta_B)}$, that is,

$$d_w = 1 + \theta_B. \tag{6.35}$$

Equation (6.35) implies that, $\langle \Delta x^2 \rangle \sim t^{1.26}$ (superdiffusion) in two dimensions, and $\langle \Delta x^2 \rangle \sim t^{0.97}$ (subdiffusion) in three dimensions. The time scale τ_s is obtained from (6.30) and is given by

$$\tau_s \sim (p - p_c)^{-\mu + \beta_B - \nu_p} \sim \xi_p^{1+\theta_B}, \tag{6.36}$$

which, for $L \ll \xi_p$, yields $\tau_s \sim L^{1+\theta_B} \sim (\alpha_L)^{1+\theta_B}$. These results were first given by Sahimi (1987). On the other hand, (6.33) tells us that $v_B \sim \xi_p^{-\theta_B}$, and therefore in the fractal regime, $v_B \sim L^{-\theta_B} \sim \langle \Delta x \rangle^{-\theta_B}$. Since $v_B = d\langle \Delta x \rangle / dt \sim \langle \Delta x \rangle^{-\theta_B}$, we obtain $\langle \Delta x \rangle \sim t^{\theta_B/(\theta_B+1)}$, and $v_B \sim t^{-\theta_B/(\theta_B+1)}$. Therefore

$$\alpha_L \sim t^{1/(\theta_B+1)}. \tag{6.37}$$

Thus, using the value of θ_B in two dimensions we obtain $\alpha_L \sim t^{0.52}$, in excellent agreement with the prediction of (6.29).

What is the implication of this result? For length scales $L \ll \xi_p$, a percolating network is fractal and macroscopically heterogeneous (see Chapter 2). A fractal structure implies the existence of long-range correlations in the rock. Although not of the same type, such correlations in a percolating structure are consistent with the results of Hewett (1986) and others that indicate that there are long-range correlations between the permeabilities of various regions of heterogeneous rocks. Thus, a simple percolation model provides a clear explanation for the observed dispersivities of heterogeneous rocks and aquifers. Scale-dependence of D_L or α_L implies that dispersion in heterogeneous rock is not diffusive, and cannot be described by the CDE. The exact form of the equation that describes dispersion in a heterogeneous rock remains an unsolved problem.

6.7 Conclusions

Percolation and network models provide a quantitative tool for modeling dispersion in microscopically disordered but macroscopically homogeneous rocks. All of the important mechanisms of dispersion can be incorporated in such models. Moreover, percolation provides a simple explanation for the observed scale-dependence of dispersivities in heterogeneous rocks, in terms of the fractal structure of the rock and the long-range correlations that such fractal structures imply.

References

Ababou, R. and Gelhar, L.W., 1990, in *Dynamics of Fluids in Hierarchical Porous Media*, edited by J.H. Cushman, San Diego, Academic.

Aris, R., 1959, *Chem. Eng. Sci.* **11**, 194.

Arya, A., Hewett, T.A., Larson, R.G., and Lake, L.W., 1988, *SPE Reservoir Engineering* **3**, 139.

Bacri, J.-C., Rakotamala, N., and Salin, D., 1987, *Phys. Rev. Lett.* **58**, 2035.

Charlaix, E., Hulin, J.P., and Plona, T.J., 1988, *J. Phys. D.* **21**, 1727.

Gist, G.A., Thompson, A.H., Katz, A.J., and Higgins, R.L., 1990, *Phys. Fluids A* **2**, 1533.

Hewett, T.A., 1986, *Society of Petroleum Engineers Paper 15386*, New Orleans, LA.

Hulin, J.P., Charlaix, E., Plona, T.J., Oger, L., and Guyon, E., 1988, *AIChE J.* **34**, 610.

Koch, D.L. and Brady, J.F., 1985, *J. Fluid Mech.* **154**, 399.

Koplik, J., Redner, S., and Wilkinson, D., 1988, *Phys. Rev. A* **37**, 2619.

Mandelbrot, B.B. and Van Ness, J.W., 1968, *SIAM Rev.* **10**, 422.

Molz, F.J., Güven, O., and Melville, J.G., 1983, *Ground Water* **21**, 715.

Neuman, S.P., 1990, *Water Resour. Res.* **26**, 887.

Philip, J.R., 1986, *Trans. Porous Media* **1**, 319.

Roux, S., Mitescu, C., Charlaix, E., and Baudet, C., 1986, *J. Phys. A* **19**, L687.

Saffman, P.G., 1959, *J. Fluid Mech.* **6**, 321.

Sahimi, M., 1987, *J. Phys. A* **20**, L1293.

Sahimi, M., 1993, *Rev. Mod. Phys.*

Sahimi, M., Davis, H.T., and Scriven, L.E., 1983, *Chem. Eng. Commun.* **23**, 329.

Sahimi, M., Hughes, B.D., Scriven, L.E., and Davis, H.T., 1986 *Chem. Eng. Sci.* **41**, 2103.

Sahimi, M. and Imdakm, A.O., 1988, *J. Phys. A* **21**, 3833.

Sudicky, E.A., Cherry, J.A., and Frind, E.O., 1985, *Water Resour. Res.* **21**, 1035.

Turner, G.A., 1959, *Chem. Eng. Sci.* **10**, 14.

7
Two-phase flow in porous media

7.0 Introduction

We now turn our attention to the flow of two immiscible fluids in a porous medium. A large number of factors can affect this phenomenon, including capillary, viscous, and gravity forces, material properties of the two fluids, e.g., their viscosities and the interfacial tension separating them, chemical and physical properties of the surface of the pores, i.e., whether or not there are surface active agents, or whether the surface is rough and fractal, the morphology of the pore space, and the wettability of the fluids. Among these the wettability deserves special attention because it has considerable influence on how the two fluids are distributed in the pore space, and how one fluid displaces another one. Thus, we first define wettability and discuss its effect on capillary pressure.

7.1 Wettability

Generally speaking, the solid surface–fluid interactions are what we call wettability. Consider, as an example, a situation in which a drop of water is placed on a surface immersed in oil. Then, a contact angle is formed that can vary anywhere from 0° to 180°. A typical situation is shown in Fig. 7.1. The three different surface tensions are related by the Young–Dupŕe equation

$$\sigma_{ow}\cos\theta = \sigma_{os} - \sigma_{ws}, \tag{7.1}$$

where σ_{ow} is the interfacial tension between oil and water, and σ_{os} and σ_{ws} are the surface tensions between oil and the solid surface, and water and the solid surface, respectively. Normally, the contact angle θ is measured through the water phase. Strictly speaking, if $\theta > 90°$, the surface is

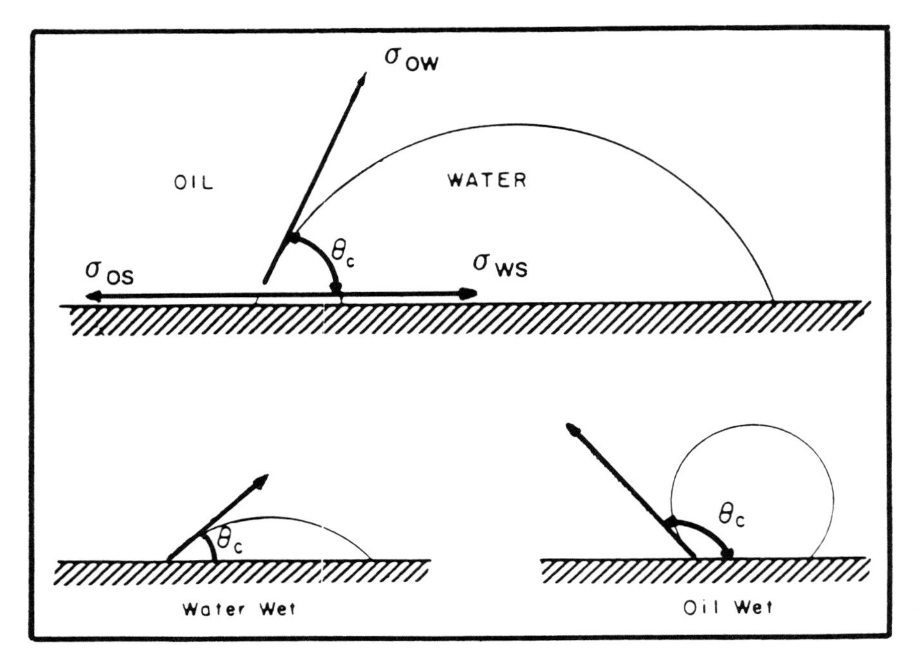

Figure 7.1 *Formation of contact angles between two liquids with different wettability and solid surface.*

preferentially oil-wet. However, in practice, $\theta < 65°$ for water-wet systems, while $105° < \theta < 180°$ for oil-wet systems. If $65° < \theta < 105°$, the system is said to be *intermediately wet*, and has no strong preference for any of the two fluids. Another important case is *mixed wettability*, in the sense that the wettability of the surface changes from pore to pore or from one portion of the surface to another. This is caused by *chemical* heterogeneity of the surface, and is actually the situation in most oil reservoirs. Many methods have been devised for measuring the wettability of a system, and Anderson (1986) has given a thorough discussion of them.

The study of moving contact lines and contact angles goes back to 1921 when Washburn proposed (3.2). This equation is invalid if the length of the tube is much longer than its diameter, and if the diameter is so small that gravity cannot have a significant effect on the shape of the moving contact line or meniscus. There is a considerable amount of experimental evidence in support of this equation. But the Washburn equation neglects the details of the surface and its effect on the moving contact line; it provides only an overall picture of what happens. Roughness, chemical heterogeneity, and other factors can make the system so complex that the Washburn equation may no longer be valid. Since there is a strong correlation between the shape of the capillary pressure curves of a medium and its transport properties, we first discuss the effect of wettability on capillary pressure.

7.2 Effect of wettability on capillary pressure

Chapter 3 introduces capillary pressure curves. They can be used to extract information about the pore size distribution of a porous medium. Before we discuss the effect of wettability on capillary pressure curves, we cover the terminology frequently used in the oil industry. In *drainage* a nonwetting fluid displaces a wetting fluid from a porous medium, while during *imbibition* a wetting phase displaces a nonwetting one. In general, we can study two kinds of processes. *Primary displacement* is the reduction of the saturation of a reference phase from 100% to the residual saturation (RS) by injection of a nonreference phase. The RS of a phase is its saturation at the percolation threshold, i.e., the point at which the fluid phase becomes disconnected, and no sample-spanning cluster of it exists. *Secondary displacement* follows primary displacement. It is reduction of the nonreference

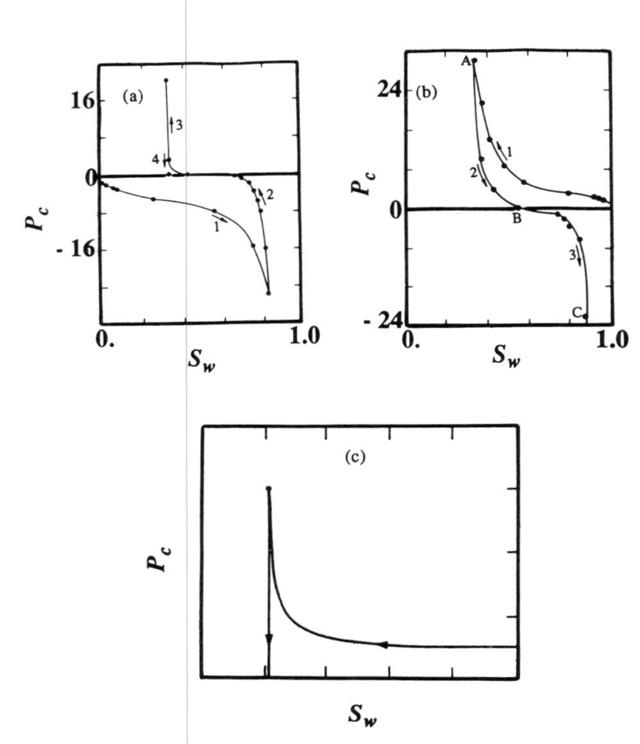

Figure 7.2 *Variations of capillary pressure P_c for various wettability regimes as a function of the water saturation S_W. The numbers on the arrows indicate the displacement mode. (a) Wetted regime, where 1 is drainage, 2 is spontaneous imbibition, 3 is forced imbibition, and 4 is secondary drainage (at the end of imbibition). (b) Nonwetted porous medium, where 1 is drainage, 2 is spontaneous imbibition, and 3 forced imbibition. (c) Mixed wettability (after Morrow and McCaffery 1978).*

phase saturation to the RS by injection of the reference phase. On the basis of the capillary pressure behavior in the sequence of primary and secondary displacements, we can identify three regimes of wettability: (i) *wetted*, in which primary displacement is drainage and secondary displacement is imbibition; (ii) *intermediate wettability*, in which primary and secondary displacements are *both* drainage, and (iii) *nonwetted*, in which primary displacement is imbibition and secondary displacement is drainage.

Figure 7.2(a) shows typical capillary pressure curves for nonwetted porous media. These curves were obtained in an oil-wet Berea sandstone treated with Drifilm (to render it strongly oil-wet). These should be compared with those shown in Fig. 6.2(b) which are typical of the wetted regime. The measurements were done on a water-wet Berea sandstone. In the primary process (drainage), denoted by 1 on the curve, oil displaces water; it ends at A. This is followed by process 2 (imbibition of water), up to point B at which the capillary pressure $P_c = 0$. Beyond B, the water has to be forced into the medium (process 3), characterized by a negative P_c, until point C is reached. Finally, Fig. 6.2(c) shows capillary pressure curves for typical intermediately wet systems. They demonstrate the value of capillary pressure curves for understanding wettability of a porous medium.

7.3 Immiscible displacements

The displacement of one fluid by another immiscible fluid is controlled and affected by factors given at the beginning of this chapter. We have discussed the effect of wettability and contact angles on capillary pressure, and later we shall discuss their effect on transport properties of the porous medium in two-phase flow. The effect of gravity is discussed in Section 7.10. Among the remaining factors, the capillary number Ca (3.5) and the mobility ratio M of the two fluids have the greatest importance. The mobility of a fluid is the ratio of the permeability and the viscosity of the fluid. Generally speaking, drainage processes are *all* percolation phenomena, whereas imbibition processes are more complex. Some of their features are related to percolation, while some of them are not. A very careful discussion and classification of displacement processes was given by Payatakes and Dias (1984). Two of such displacement processes that are related to percolation are as follows.

7.3.1 Quasistatic imbibition

In this process at any given step only one pore is invaded by the displacing fluid. This can be done by adjusting the backpressure so that the *narrowest* pore throat is invaded, while the interface at other larger throats remains essentially motionless. Since even the largest pore throats are smaller than the pore body to which they are connected, once a pore body is invaded, all of the throats that are connected to it are also invaded. As soon as the interface enters such throats, the smallest pore body that is connected to those throats is invaded and so on. Thus, at any given step of the displacement, the smallest accessible pore body (in the percolation sense) is invaded.

When the displacing fluid forms a sample-spanning cluster of invaded pore bodies (and the associated pore throats), a *breakthrough* is achieved. Just before breakthrough, the displaced fluid, in a three-dimensional porous medium, is mostly connected. As the displacement proceeds, small blobs of the displaced fluid are formed and become trapped. At the end of this process, we may end up with a large number of isolated blobs whose total saturation depends on the morphology of the pore space. In unconsolidated media it varies between 0.14 and 0.2, whereas in consolidated rocks it is anywhere between 0.4 and 0.8.

Experimental data indicate that the size distribution of the blobs, when expressed in terms of the number n_{sb} of pore bodies they occupy, follow a power law

$$n_{sb} \sim s^{-\beta_b}, \tag{7.2}$$

reminiscent of (2.14), the scaling law for the number of finite percolation clusters of s sites. Indeed, an appropriate percolation model can be devised to model such a quasistatic displacement.

7.3.2 Dynamic invasion at constant flow rates

The driving force for this process is an applied pressure drop, and the role of capillary forces is of secondary importance. If $M > 1$, we will have a complex unstable displacement (see Sahimi 1993 for a more complete discussion). Thus, we only consider the case $M < 1$. Suppose that the capillary pressure is negligible compared to the applied pressure. If so, at any stage of the displacement we will have *several* advancing interfaces in as many pores. Because the driving force is the applied pressure, the microscopic interfaces choose the *largest* accessible pore throats to minimize the resistance. Thus, the structure of the sample-spanning cluster of the displacing fluid is very different from the structure of the imbibition process; it resembles a drainage process. But this does not necessarily mean that smaller throats will not be selected; local pressures are also important and can cause the invasion of smaller throats by the advancing fluid.

At the end of these two processes, the displaced fluid exists only in the form of isolated blobs of finite sizes that cannot be displaced by secondary drainage or imbibition. The distribution of the blobs is similar to the distribution of the clusters at or below the percolation threshold, where each cluster represents an isolated blob.

7.4 Relative permeabilities to two-phase flow

There are essentially two classes of models of two-phase flow and displacement in porous media. One class relies on continuum equations, averaged over suitably defined representative volume. This is the classical engineering approach; the literature on this class of models is enormous and I refer the reader to Marle (1981). The other class contains discrete or statistical models, some of which are discussed in the context of single-phase flow and dispersion in Chapters 5 and 6. The literature on this class of models has also grown dramatically in the last few years. It would be impossible to discuss everything so the reader is referred to Sahimi (1993) for a more complete discussion. We only discuss percolation-based models.

Modeling two-phase flow in porous media is usually done by using the concept of *relative permeability* (RP). Starting with the continuity and Stokes' equations for each phase (Chapter 5), and using the appropriate boundary conditions, we can derive the following equations for the average flow velocities and saturations S_β and S_γ of the phases β and γ at time t

$$\mathbf{v}_\beta = -\frac{\mathbf{K}_\beta}{\eta_\beta} \cdot (\nabla \langle p_\beta \rangle^\beta - p_\beta \mathbf{g}) + \mathbf{K}_{\beta\gamma} \cdot \mathbf{v}_\gamma, \tag{7.3}$$

$$\frac{\partial S_\beta}{\partial t} + \nabla \cdot \mathbf{v}_\beta = 0, \tag{7.4}$$

$$\mathbf{v}_\gamma = -\frac{\mathbf{K}_\gamma}{\eta_\gamma} \cdot (\nabla \langle p_\gamma \rangle^\gamma - p_\gamma \mathbf{g}) + \mathbf{K}_{\gamma\beta} \cdot \mathbf{v}_\beta, \tag{7.5}$$

$$\frac{\partial S_\gamma}{\partial t} + \nabla \cdot \mathbf{v}_\gamma = 0. \tag{7.6}$$

The right sides of (7.3) and (7.5) contain two terms. The first term is the usual Darcy's law, written for each phase, while the second term couples the two phases. Equations (7.3)–(7.6) are valid if $Ca \ll 1$, and if the process is in a quasisteady state condition. However, the coupling terms in (7.3) and (7.5) are not significant unless $\eta_\beta \simeq \eta_\gamma$ and, for this reason, such cross terms are generally neglected. The *phase* permeabilities $\mathbf{K}_\beta$ and $\mathbf{K}_\gamma$ are supposed to be known, but in practice we calculate them using the relation

$$\mathbf{K}_\beta = \mathbf{K} k_{r\beta}, \tag{7.7}$$

where $\mathbf{K}$ is the permeability tensor, and $k_{r\beta}$ is the RP to the β-phase (with a similar equation for the γ-phase). A major problem in any two-phase flow in porous media is the prediction of the RPs. Unlike the absolute permeability, $k_{r\beta}$ has been found to depend on many parameters, including saturation and saturation histories of the fluids, the pore space morphology, the wetting characteristics of the fluids, Ca, and sometimes M. Moreover, 40 years ago it was recognized (Richardson *et al.* 1952) that the RP to a phase typically becomes very small or altogether negligible when its saturation is less than a critical value, which is distinctly above zero. This is of course the signature of a percolation problem discussed below.

There are many ways of measuring RPs and phase saturations (see Anderson 1987). In a routine method, the porous medium is initially filled with the wetting phase and a mixture of wetting and nonwetting fluids is injected into the medium at a constant flow rate. When the steady state has been reached, the pressure drop across the medium is recorded. From the knowledge of the flow rate and pressure drop, the phase permeabilities k_β and k_γ are calculated using (7.3) and (7.5). The simplest way of measuring the saturation is by weighing the sample before and after injection. Since the absolute permeability of the medium is already known (see Chapter 5), the RP to the wetting phase at this particular saturation is calculated. In the next

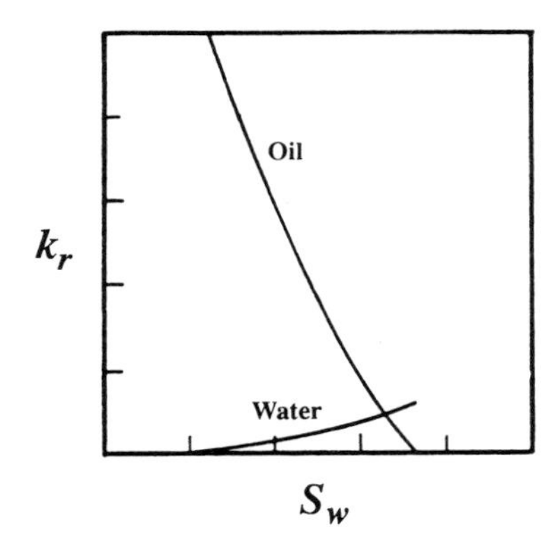

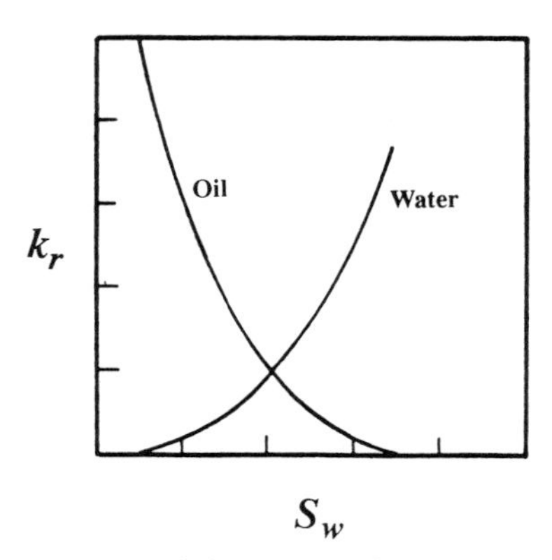

Figure 7.3 *Typical relative permeability curves for a water-wet porous medium (top), and for an oil-wet porous medium (bottom).*

stage, the injected amount of nonwetting fluid is increased, and the procedure is repeated. In this way, the RP to the wetting phase is obtained. By reversing the procedure we can obtain the RP to the nonwetting phase.

7.5 Effect of wettability on relative permeability

Similar to capillary pressure, wettability and contact angles also affect significantly the RPs. Figure 7.3(a) shows typical oil–water RP curves for a strongly water-wet system, while Fig. 7.3(b) shows the same for a strongly oil-wet porous medium. While the difference between the two cases for the RP to the oil phase is not very large, there is a dramatic difference between the RPs to the water phase in the two cases. Normally, if a system is strongly water-wet, there is little or no hysteresis in the RPs to the water phase, and this can be clearly seen in any experiment. Thus, RP curves can also be characteristic of the wettability of a porous medium.

7.6 Percolation models of capillary-controlled two-phase flow and displacement

We now discuss percolation models of two-phase flow and displacement in porous media which are, strictly speaking, applicable only when *Ca* is very small. There are also a few network models for the case when *Ca* is finite (in which case both capillary and viscous forces are relevant). But such models are not generally related to percolation and, therefore, are not discussed here. The reader can consult Sahimi (1993) for a review of this class of models.

7.6.1 Random percolation models

At the outset we should point out that the fundamental assumption in *all* percolation models of two-phase flow is that the bond occupation probability p, defined in Chapter 2, is proportional to the capillary pressure. Without this assumption, it would be difficult to make a one-to-one correspondence between a percolation model and the two-phase flow problem. Although in some percolation models, such as invasion percolation, the occupation probability is not used, an analog of this quantity can be defined.

Fatt (1956) was the first to use a network model for simulating two-phase flow in a porous medium, and Melrose and Brandner (1974) were the first to suggest that percolation may be useful for describing two-phase flow in porous media, but the first random percolation model of two-phase flow in porous media was suggested by Larson *et al.* (1977), with the full details of their work given in Larson *et al.* (1981a,b). Larson *et al.* (1981a) proposed a model for drainage in which the porous medium was represented as a cubic network of bonds and sites with distributed sizes. It was assumed that a bond next to the interface between the displaced and displacing fluids is penetrated by the displacing fluid if the capillary pressure at that point exceeds a critical value. This implies that the radius of that bond has to exceed a critical radius r_{min} defined by (3.6). All bonds that are connected to the displacing fluid by a path of bonds with effective radius larger than r_{min} are considered as *accessible*; the accessibility is defined in the sense of percolation discussed in Chapter 2. It was also assumed that any accessible bond whose radius is at least as large as r_{min} is filled with the nonwetting fluid. This is of course not true, since an interface which starts at one external face of a porous medium has to travel a certain path of eligible bonds before it reaches an accessible and potentially eligible bond.

In another paper, Larson *et al.* (1981b) proposed a percolation model of imbibition in order to calculate the residual nonwetting phase saturation and its dependence on Ca. They modeled the creation of isolated blobs of the nonwetting phase by a random site percolation (Chapter 2). At the site percolation threshold of the network, they calculated the fraction $\hat{g}(s)$ of the occupied sites that are in clusters of length s in the direction of flow, and argued that this represents the desired blob size distribution. To calculate the RS of the nonwetting phase S_{rnw}, they assumed that once a blob is mobilized, it is permanently displaced. This is not always true, because a blob can get trapped again, can join another blob and create a larger one, and so on.

The fundamental assumption behind the Larson *et al.* models is that pore-level events are controlled by capillary forces. It is possible to devise simple scaling arguments to estimate the Ca for which this assumption is valid. The capillary pressure across the interface is proportional to

$$P_c \sim \frac{\sigma_{ow} \cos\theta}{d_g}, \tag{7.8}$$

where d_g is a typical grain size. On the other hand, the viscous pressure drop is proportional to

$$P_{vis} \sim \frac{\eta_w v d_g}{k}, \tag{7.9}$$

where η_w is the viscosity of the wetting phase. Therefore,

$$\frac{P_{vis}}{P_c} \sim \frac{Ca}{k_d} \tag{7.10}$$

where $k_d = k/d_g^2$ is a dimensionless permeability, which is small (of the order of 10^{-3} or smaller), because k is controlled by the narrowest throats in the medium. It follows that for capillary-controlled displacements, we must have $Ca \ll 1$, and in practice we have $Ca \sim 10^{-5} - 10^{-6}$. In many experiments it has been observed that S_{rnw} is constant for $Ca < Ca_c$, where Ca_c is the critical value of Ca for capillary-controlled displacement, and S_{rnw}

decreases only when $Ca > Ca_c$. Larson *et al.* (1981b) compiled a wide variety of experimental data that supported this; the significance of their work is that it was the *first* attempt for using percolation ideas for quantitative modeling of two-phase flow in porous media.

Heiba *et al.* (1982) further developed these ideas and applied them to the calculation of RPs. Full details are given by Heiba *et al.* (1992). Heiba *et al.* (1982) distinguished between pore throats that are *allowed* to a phase, and those that are actually occupied by the phase. Then, given a pore size distribution of the pore space, they calculated the pore size distribution of the allowed and occupied pores. Consider, for example, a displacement process in which one fluid is strongly wetting, while the other one is completely nonwetting. Then, according to their model, during primary drainage the pore size distribution of the pores occupied by the displacing phase is given by (3.7), since the *largest* throats are occupied by the displacing fluid. During imbibition the pore size distribution of the pores occupied by the displacing phase is given by (3.9), because the *smallest* pores are occupied by this phase. We can, in a similar fashion, derive expressions for the pore size distribution of the pores occupied by the displacing and displaced fluids during secondary imbibition and drainage. Once these pore size distributions are determined, calculating the permeability or hydraulic conductivity of each fluid phase (and, therefore, the RP) reduces to a problem of percolation conductivity, because when we calculate the permeability to a given phase, the flow conductance (or effective radii) of the bonds occupied by the other phase can be set to zero, since the two phases are immiscible. Therefore, any of the methods discussed in Chapter 5 can be used for calculating the RPs (see, for example, Sahimi 1988, Heiba *et al.* 1992). Figure 7.4 shows the predictions of this model using a cubic network; all important aspects of the experimental data (Fig. 7.3) are reproduced by the model. Permeability and RPs are controlled by the pore throats, and therefore pore bodies do not play any important role. However, their volumes have to be taken into account when calculating the phase saturations. This random percolation model can also be extended for calculating the RPs for the intermediate and mixed wettability conditions; see Heiba *et al.* (1983).

7.6.2 Invasion percolation

This model was proposed by Lenormand and Bories (1980) and Chandler *et al.* (1982). In this model the network is initially filled with the fluid to be displaced, often called the *defender*. To each site of the network is assigned a random number uniformly distributed in [0,1]. Then, the displacing fluid, or the *invader*, is injected into the medium and displaces the defender at each time step by choosing the site on the interface that has the smallest random number. If we interpret the random numbers as the resistance that

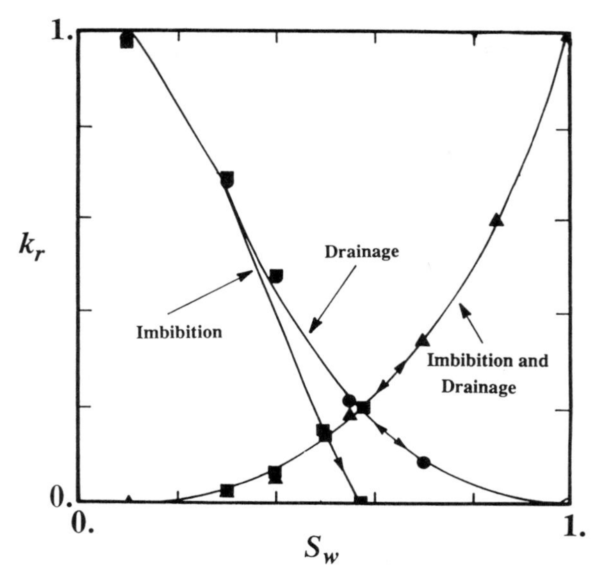

Figure 7.4 *Relative permeability k_r versus wetting phase saturation S_w for a strongly wetting, completely nonwetting porous medium, as predicted by the random percolation model using a simple cubic network.*

the sites offer to the invader, then choosing the site with the smallest random number is equivalent to selecting a pore with the largest size, hence this model simulates a drainage process. A slightly more tedius procedure can be used for working with bonds instead of sites. Invasion percolation is a dynamic cluster growth process; random percolation is quasistatic.

Two versions of this model have been studied. In one model, the defender is an incompressible fluid. Thus, if a blob of it is surrounded by the invader, it cannot be penetrated and is *trapped*. Computer simulations of Wilkinson and Willemsen (1983) indicated that in two dimensions the fractal dimension of the invader cluster is about 1.82, smaller than that of random percolation which is $91/48 \simeq 1.896$. But no significant difference was observed in three dimensions. In the second model, the defender is assumed to be compressible, so that even if a blob of it is surrounded by the invader, it can still be penetrated. There is a close connection between this version of invasion percolation and the random percolation model. This was demonstrated by Wilkinson and Barsony (1984) who showed that the exponent $\Delta_p = \beta_p + \gamma_p$ (Chapter 2) is the same for both random and invasion percolation models in both two and three dimensions. Therefore, this form of invasion percolation is in the universality class of random percolation.

From a fundamental point of view, invasion percolation is definitely a more appropriate model of immiscible displacements than the random percolation model. The most obvious reason for this is the fact that there is a well-defined interface that starts from one side of the system and displaces the defender in a systematic and realistic way. Thus, the history and the sequence of invading pores, according to a physical rule, have been naturally built into the model.

Let us mention some experimental evidence in support of the invasion percolation model of two-phase flows; for a more complete discussion see Sahimi (1993). Lenormand and Zarcone (1985a) displaced oil (the wetting fluid) by air (the nonwetting fluid) in a large and transparent two-dimensional etched network and obtained a fractal dimension of about 1.82, consistent with two-dimensional computer simulations of invasion percolation with trapping. Shaw (1987) showed that if a porous medium, filled with water, is dried by hot air, the dried pores (those filled with air) form an invasion percolation cluster with the same fractal dimensionality as that found in computer simulations. Lenormand and Zarcone (1985b) used two-dimensional etched networks and a variety of wetting and nonwetting fluids (oil, different water–sucrose solutions, air), and showed that their drainage experiments are all completely consistent with an invasion percolation description of this phenomenon.

Invasion percolation has some interesting and unusual properties. For example, Furuberg *et al.* (1988) studied the probability $P(r,t)$ (where $r = |\mathbf{r}|$) that a site, a distance r from the injection point, is invaded at time t. They found that a dynamic scaling governs $P(r,t)$

$$P(r,t) \sim r^{-1} f(r^{D_c}/t), \tag{7.11}$$

where D_c is the fractal dimension of the invasion percolation cluster, and $f(u)$ is a scaling function with the unusual property that, $f(u) \sim u^{a_1} (u \ll 1)$ *and*, $f(u) \sim u^{-a_2} (u \gg 1)$, i.e., $f(u)$ vanishes at *both* ends. This dynamic scaling implies that the most probable growth takes place at $r \sim t^{1/D_c}$. The reason for this unusual limiting behavior of $f(u)$ is that at time t, most of the region within the distance r has already been invaded, and new sites close to the interface that can be invaded are rare. Roux and Guyon (1989) argued that the exponents a_1 and a_2 are given by, $a_1 = 1$, and $a_2 = \tau_p + \sigma_p - D_h/D_c - 1$, where τ_p, σ_p, and D_c are the usual percolation exponents and fractal dimension (Chapter 2), and D_h is the fractal dimension of the *hull* (or surface) of percolation clusters, where $D_h(d=2) = 1 + 1/\nu_p = 7/4$, and $D_h(d=3) \simeq D_c$.

7.6.3 Random percolation with trapping

Computer simulations of invasion percolation with trapping by Wilkinson and Willemsen (1983) in two dimensions indicated that the fractal dimension of the invasion cluster is about 1.82, as compared with $D_c = 91/48 \simeq 1.896$ for random percolation (Chapter 2). The experiments of

Lenormand and Zarcone (1985a) supported this. Random percolation with trapping was first developed by Sahimi and Tsotsis (1985) to model a completely different phenomenon, namely, catalytic pore plugging of porous media, which will be discussed in Chapter 8. It was further investigated by Dias and Wilkinson (1986) as a model of two-phase flow in porous media, whose results indicated that the scaling properties of this model in *both* two and three dimensions are the same as those of the usual random percolation discussed in Chapter 2.

For calculating the RPs, the predictions of most percolation models are very similar. But the random percolation model has the advantage that it enables us to derive analytical formulae for the pore size distributions of various phases during both imbibition and drainage, discussed above and in Chapter 3. We can employ these formulae with the methods discussed in Chapter 5 to estimate the RP to both phases during imbibition and drainage; see Blunt *et al.* (1992) and Sahimi (1993) for a more complete discussion of this.

7.7 Crossover from percolation to compact displacement

Although we discussed RP curves for *both* imbibition and drainage in terms of a percolation model, there are qualitative differences between the two that need to be discussed. A clue to these differences is already evident in the RP curves (Fig. 7.4). The RP to the nonwetting phase during primary imbibition by a strongly wetting fluid vanishes only at $S_{rnw} = 0$, i.e., the nonwetting phase is completely expelled from the system and the wetting phase fills the system. However, during drainage by a completely nonwetting fluid the RP to the wetting phase vanishes at a *finite* value of S_{rw}, i.e., the nonwetting phase does not fill the porous medium, and a fractal percolation cluster is formed.

A definitive study of this phenomenon was made by Cieplak and Robbins (1990), who represented the porous medium by a two- dimensional array of disks with random radii, where the underlying lattice was either a triangular or a square network. The limit of low *Ca* was considered, and the displacement dynamics were modeled as a stepwise process where each unstable section of the interface moved to the next stable or nearly stable configuration. Their simulations showed that there are three basic types of instability and corresponding growth mechanisms. (i) *Burst* happens when, at a given P_c, no stable arc connects two disks and, therefore, the interface simply jumps forward to connect to the nearest disk. (ii) *Touch* happens when an arc that connects two disks intersects another disk at a wrong contact angle θ. In this case, the interface connects to this third disk. (iii) *Overlap* happens when two nearby arcs overlap. There is no need for

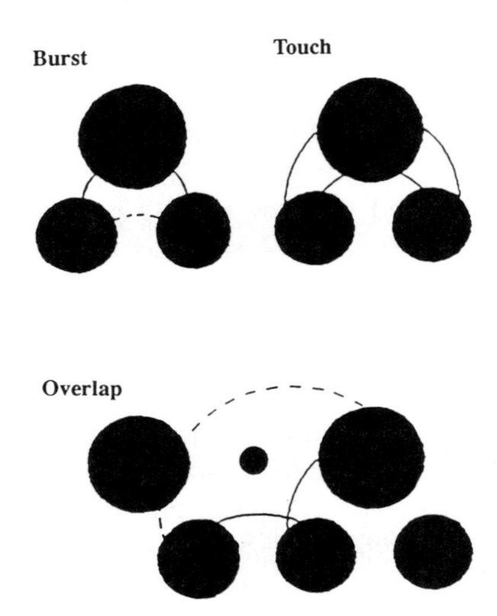

Figure 7.5 *Three types of instability and growth that occur during an immiscible displacement (after Cieplak and Robbins 1990).*

the disk to which both arcs are connected, and it can be removed from the interface. Figure 7.5 illustrates these three growth mechanisms.

To simulate the growth of the interface, a capillary pressure P_c is fixed and the stable arcs are found. If instabilities are found, local changes are made to remove them. Then P_c is increased by a small amount, the interface is advanced, possible instabilities are removed again, and so on. As in percolation with trapping, if the invading fluid surrounds a blob of the displaced fluid, the blob is kept intact for the rest of the simulation. If all disks have the same radius, the resulting patterns are very regular and faceted, which also preserves the symmetry of the lattice. However, when the radii of the disks are randomly distributed, then the behavior of the system depends on the contact angle. To quantify the effect of θ, an interface width w is used. If $\theta \simeq 180°$ (drainage), then the phenomenon is a percolation process with the same fractal structure and dimension as invasion percolation, and w is of the order of the pore size. As θ decreases, w increases, and the cluster of invading fluid becomes more compact; see Fig. 7.6. Finally, at a critical contact angle θ_c, the width w *diverges* according to a power law

$$w \sim (\theta - \theta_c)^{-\nu_\theta}, \qquad (7.12)$$

where $\nu_\theta \simeq 2.3$ in the two-dimensional simulations. The critical angle θ_c was found to depend on the porosity ϕ of the system, for example, $\theta_c \simeq 29°$ for $\phi = 0.322$, and $\theta_c \simeq 69°$ for $\phi = 0.73$. The exponent ν_θ was found to be universal. The compactness of the cluster for $\theta < \theta_c$ is consistent with the imbibition picture discussed above.

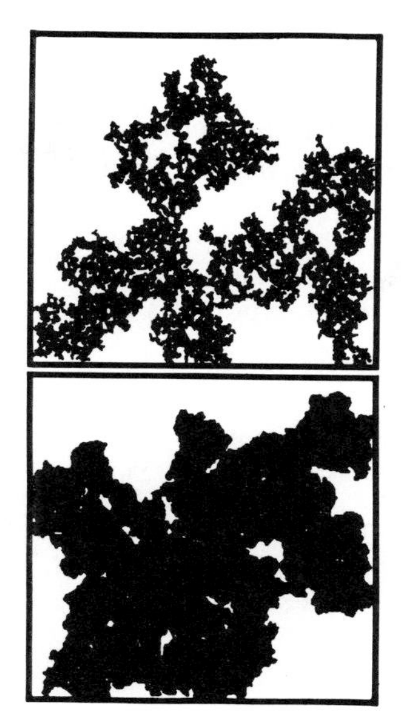

Figure 7.6 *Percolating displacement patterns for* $\theta = 179°$ *(top) which is a drainage process, and* $\theta = 58°$ *(bottom) which is an imbibition process (after Cieplak and Robbins 1990).*

The divergence of w at θ_c is clearly due to the transition from a percolation-like (fractal-like) displacement to a compact one. For large θ, growth occurs mainly by burst, which is similar to invasion percolation, and the growth pattern is independent of θ. However, as $\theta \to \theta_c$, the overlap and touch phenomena become more important, and the interface is unstable for almost any configuration of the local geometry. Thus, the growth pattern changes, hence w diverges.

7.8 Roughening and pinning of fluid interfaces: dynamic scaling and directed percolation

Although during imbibition the cluster of the invading fluid is compact, the capillary forces lead to random local pinning of the interface, resulting in a rough interface with a self-affine (anisotropic) fractal structure. This was demonstrated by the experiments of Rubio *et al.* (1989) and Horváth *et al.* (1991a). The only difference between the two experiments was in the wetting fluid used. Rubio *et al.* (1989) performed their experiments in a thin (essentially two-dimensional) porous medium made of tightly packed clean glass beads of various diameters. Water was injected into the porous medium to displace the air in the system. The motion of the interface was recorded and digitized with high resolution. The water region was always compact, but the interface between water and air was rough.

The roughness of the interface can be characterized by the width $w(L)$, which is defined as $w(L) = \langle (h(x) - \langle h \rangle_L)^2 \rangle^{1/2}$, where h is the height of the interface at position x, and $\langle h \rangle_L$ is its average over a horizontal segment of length L. According to the scaling theory of Family and Vicsek (1985) for growing rough surfaces one has the following scaling form at time t

$$h(x) - \langle h \rangle_L \sim t^{\beta} f(x/t^{\beta/\alpha}), \tag{7.13}$$

where α and β are two critical exponents that satisfy the following scaling relation

$$\alpha + \frac{\alpha}{\beta} = 2 \tag{7.14}$$

The scaling function $f(y)$ has the properties that, $|f(u)| < c$ for $u \gg 1$, and $f(u) \sim L^{\alpha} f(Lu)$ for $u \ll 1$, where c is a constant. Given (7.13), it is easy to see that

$$w(L, t) \sim t^{\beta} g(t/L^{\alpha/\beta}), \tag{7.15}$$

where $g(u)$ is another scaling function, and therefore

$$w(L, \infty) \sim L^{\alpha}. \tag{7.16}$$

Note that $w(L, t)$ is a measure of the correlation length along the direction of growth. A variety of surface growth models and the resulting dynamical scaling can be described by the stochastic differential equation proposed by Kardar, Parisi, and Zhang (1986)

$$\frac{\partial h}{\partial t} = \sigma_i \nabla_T^2 h + \frac{v_g}{2} |\nabla h|^2 + \mathcal{N}(\mathbf{r}, t), \tag{7.17}$$

where σ_i corresponds to the effect of surface tension, v_g is the growth velocity perpendicular to the interface, and $\mathcal{N}$ is a noise term that somehow represents disorder. Kardar *et al.* (1986) considered the case in which the noise was assumed to be Gaussian with the correlation

$$\langle \mathcal{N}(\mathbf{r}, t) \mathcal{N}(\mathbf{r}', t) \rangle = 2A\delta(\mathbf{r} - \mathbf{r}', t - t'), \tag{7.18}$$

where A is the amplitude of the noise. For this model it has been proposed that (Kim and Kosterlitz 1989, Hentschel and Family 1991)

$$\alpha = \frac{2}{d+2}, \quad \beta = \frac{1}{d+1}, \tag{7.19}$$

for a d-dimensional system. A related stochastic differential equation was proposed by Koplik and Levine (1985)

$$\frac{\partial h}{\partial t} = \sigma_i \nabla_T^2 h + v_g + A\mathcal{N}(\mathbf{r}, h), \tag{7.20}$$

so that the term representing the noise is more complex than that of Kardar *et al.* (1986). It is now easy to see why a pinning transition occurs. Consider (7.20) in zero transverse dimension

$$\frac{\partial h}{\partial t} = v_g + A\mathcal{N}(h). \tag{7.21}$$

If $v_g > A\mathcal{N}_{max}$, where $\mathcal{N}_{max}$ is the maximum value of $\mathcal{N}$, then $\partial h/\partial t > 0$ and the interface always moves with a velocity fluctuating around v_g. If, however, $v_g < A\mathcal{N}_{max}$, the interface will eventually arrive at a point where, $v_g + A\mathcal{N} = 0$, and get pinned down. Therefore, there has to be a pinning transition at some finite value of A, if we fix v_g.

Let us now go back to the experiments of Rubio *et al.* (1989) and Horváth *et al.* (1991a). Rubio *et al.* (1989) found that $\alpha \simeq 0.73$, significantly different from $\alpha = 1/2$, predicted by (7.19). Horváth *et al.* (1990) reanalyzed the Rubio *et al.* data and obtained, $\alpha \simeq 0.91$, even larger than $\alpha = 1/2$, while for their own experiments they obtained, $\alpha \simeq 0.81$, and $\beta \simeq 0.65$. Although their α and β satisfy (7.14), they are still significantly different from the predictions of (7.19).

How can we explain these beautiful, but unexpected results? A key idea may be that a disordered porous medium generates noise that is not necessarily Gaussian, as assumed by Kardar *et al.* (1986). Several models with various forms of noise have been introduced, but the model proposed by Zhang (1990) in which the distribution of the noise amplitude is of power law form

$$H(A) \sim A^{-(\delta+1)} \tag{7.22}$$

is interesting and may be relevant, because Horváth *et al.* (1991b) showed that with such a distribution, the above experimental data can be fitted if $\delta \simeq 2.7$. However, the origin of this power law noise, and why it should be present in a porous medium, is not clear yet.

A possible connection between the roughening and pinning of the interface in imbibition and a variant of percolation, called *directed percolation*, was proposed by Buldyrev *et al.* (1992) and Tang and Leschhorn (1992). In directed percolation (for a review see Kinzel 1983), the bonds of a network are directed and diode-like. Transport along such bonds is allowed in only one direction. If the direction of the external potential is reversed, no transport in the diode-like bonds takes place. This induces a macroscopic anisotropy, so we need *two* correlation lengths for characterizing the system. One, ξ_L, is for the longitudinal (external potential) direction, while the other, ξ_T, is for the transverse (perpendicular to the external potential) direction. The percolation thresholds p_{cd} of directed networks are much *larger* than those of ordinary percolation discussed in Chapter 2. Near p_{cd} we have

$$\xi_L \sim (p - p_{cd})^{-\nu_L}, \tag{7.23}$$

$$\xi_T \sim (p - p_{cd})^{-\nu_T}. \tag{7.24}$$

Buldyrev *et al.* (1992) carried out an interesting experiment in which they formed an interface by dipping paper (a two-dimensional system) into a fluid, and allowing it to invade the paper. They and Tang and Leschhorn (1992) argued that, $w(L) \sim \xi_T$, $L \sim \xi_L$, i.e., ξ_T is the characteristic length scale for the width, while ξ_L sets characteristic scales for *both* the distance parallel to the interface and the time. Therefore

$$\alpha = \frac{\nu_T}{\nu_L}, \tag{7.25}$$

which for $d = 2$ with $\nu_L \simeq 1.73$ and $\nu_T \simeq 1.1$ yields, $\alpha \simeq 0.63$, in perfect agreement with their measurement. But it is not clear that their elegant experiment is relevant to imbibition in real rocks and porous media. Moreover, their value of α is lower than that of the above experiments, so the problem is not solved yet.

7.9 Finite size effects on capillary pressure and relative permeability

Our theoretical discussion so far has been limited to systems that are essentially of infinite extent. If the system is of finite size, the dependence of the macroscopic properties on the size L of the system can be investigated using finite-size scaling discussed in Chapter 2. Let us now discuss the effect of the size of a porous medium on its capillary pressure and RPs. Thompson *et al.* (1987) measured the electrical resistance of sandstones during mercury injection, and found that the resistance decreases (the permeability or relative permeability increases) in a stepwise manner; see Fig. 7.7. The observed steps were irreversible in that small hysteresis loops did not retrace the steps and the steps were not reproduced on successive injections. When the number $N_{\Delta R}$ of resistance steps larger than ΔR was plotted versus ΔR, a power law relation was found

$$N_{\Delta R} \sim (\Delta R)^{\lambda_R}. \tag{7.26}$$

λ_R, which was found to vary between 0.57 and 0.81, presumably depends on the strength of the competition between capillary and gravitational forces.

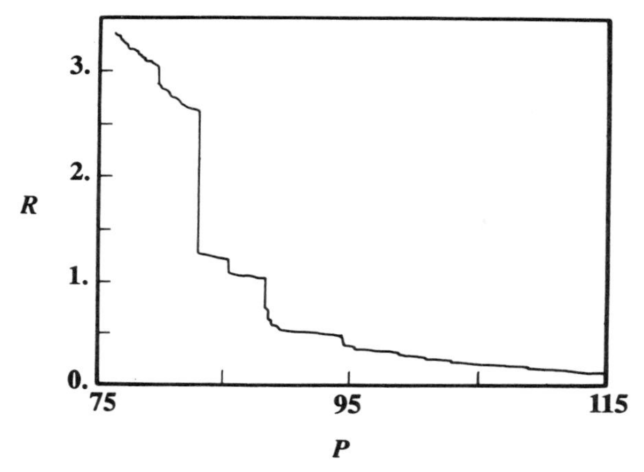

Figure 7.7 *Resistance R of a sample sandstone as a function of the injection pressure P during mercury porosimetry (after Thompson et al. 1987).*

$\lambda_R \simeq 0.57$ signifies the limit in which gravitational forces are absent, while $\lambda_R \simeq 0.81$ represents the limit in which gravitational forces are prominent. The reason for this stepwise decrease in the sample resistance is that in a *finite sample*, penetration of any pore by mercury causes a *finite* change in the resistance, but as the sample size increases, the size of the change decreases; for a very large sample the step size should vanish and the resistance curve should become continuous. Using a percolation model, Roux and Wilkinson (1988) showed that for a three- dimensional porous medium of size L

$$N_{\Delta R} \sim L^{3(\mu - \nu_p)/(\mu + 3\nu_p)}(\Delta R)^{-3\nu_p/(\mu + 3\nu_p)}, \tag{7.27}$$

so that, $\lambda_R = 3\nu_p/(\mu + 3\nu_p) \simeq 0.57$, which agrees well with the experimental result in the absence of gravity. Thus, while sample size effects are important, the stepwise decrease in the resistance of the sample during mercury injection is still consistent with a percolation description of this process, discussed in Chapter 3, and with a two-phase flow description, discussed here.

7.10 Immiscible displacements under the influence of gravity: gradient percolation

So far, we have neglected the effect of gravity on immiscible displacements. But in three-dimensional porous media, the hydrostatic component of pressure adds to the applied pressure, and this creates a vertical gradient in the effective injection pressure. Because of this gradient, the fraction of accessible pores decreases with the height of the system. This effect was not taken into account in the percolation models described above. However, a modified percolation model developed by Wilkinson (1984) succeeded in taking into account such effects.

The competition between gravity and capillary forces is usually expressed through the Bond number Bo which is defined as

$$\mathrm{Bo} = \frac{\Delta\rho g l^2}{\sigma_i} \tag{7.28}$$

where $\Delta\rho$ is the density difference between the two fluids and g is the gravity. Wilkinson (1984) showed that in an immiscible displacement under gravity, the corresponding correlation length ξ_g does *not* diverge (unlike random and invasion percolation, which have a diverging correlation length ξ_p), but it reaches a maximum value, given by

$$\xi_g \sim \mathrm{Bo}^{-\nu_p/(1+\nu_p)}, \tag{7.29}$$

so that $\xi_g \sim \mathrm{Bo}^{-0.47}$ in three dimensions and $\xi_g \sim \mathrm{Bo}^{-4/7}$ in two dimensions. In three dimensions, there is a transition region where both phases can percolate, and the width w of this region is given by

$$w \sim \mathrm{Bo}^{-1}. \tag{7.30}$$

Similar results were obtained by Sapoval *et al.* (1985) in the context of *gradient percolation*, a model in which a gradient G for the occupation probability p is imposed on one direction of the network. They used arguments similar to Wilkinson to show that

$$\xi_g \sim G^{-\nu_p/(1+\nu_p)}, \tag{7.31}$$

which is completely similar to (7.29). Birovljev *et al.* (1991) used a transparent two-dimensional porous medium consisting of a monolayer of 1 mm glass beads placed at random and sandwiched between two plates. The medium was filled with a glycerol–water mixture and was displaced by air invading the system at one end. They obtained, $\xi_g \sim \mathrm{Bo}^{-0.57}$, where the exponent 0.57 agrees completely with the theoretical prediction, $\nu_p/(1+\nu_p) = 4/7 \simeq 0.57$.

Wilkinson (1984) also derived an important result regarding the effect of gravity on the residual oil saturation (ROS). He showed by a scaling argument that the difference $S_{or} - S_{or}^0$, where S_{or} is the ROS for $\mathrm{Bo} \neq 0$ and S_{or}^0 is the corresponding value when $\mathrm{Bo} = 0$, is given by

$$S_{or} - S_{or}^0 \sim \mathrm{Bo}^{\lambda_B}, \tag{7.32}$$

where $\lambda_B = (1 + \beta_p)/(1 + \nu_p)$, ν_p and β_p are the usual percolation exponents (Chapter 2), so that, $S_{or} - S_{or}^0 \sim \mathrm{Bo}^{0.74}$, for a three-dimensional porous medium. Wilkinson (1984) also proposed a simple percolation model for simulating displacements under the influence of gravity, in which a bias, linearly proportional to the height of the interface, is added to the random numbers that are assigned to the sites in the invasion percolation. The simulation proceeds as in invasion percolation by invading those sites in the vicinity of the interface that have the smallest numbers.

7.11 Scaling laws for relative permeability and dispersion coefficients

We can derive scaling laws for capillary pressures, RPs, and dispersion coefficients near the RSs, which are the analogs of percolation thresholds. The main problem in deriving such scaling laws is relating the saturations S_w of the wetting phase and S_{nw} of the nonwetting phase (and their residual values S_{rw} and S_{rnw}) to the occupation probability p and the percolation threshold of the system. Let us briefly discuss these for RPs and the dispersion coefficient D_L. More details are given by Wilkinson (1986) and Sahimi and Imdakm (1988).

During drainage we are concerned with the point where the displacing nonwetting fluid first percolates. Since in three dimensions the trapping of the wetting phase is not important, S_{nw} is proportional to, L^{D_c}/L^3, if $L \ll \xi_p$. For $L \gg \xi_p$, we replace L with ξ_p, so that, $S_{nw} \sim \xi_p^{-\beta_p/\nu_p}$, and, therefore

$$S_{nw} \sim (p - p_c)^{\beta_p}. \tag{7.33}$$

On the other hand, during imbibition we have

$$S_{nw} - S_{rnw} \sim \int_{p_c}^{p} X^A dp \sim (p - p_c)^{1+\beta_p}. \tag{7.34}$$

Given these two equations, it is easy to derive the scaling laws for RPs and dispersion coefficients. Thus, the RP k_{rnw} to the nonwetting phase during drainage obeys, $k_{rnw} \sim (p - p_c)^{\mu}$, or in view of (7.33)

$$k_{rnw} \sim (S_{nw})^{\mu/\beta_p}, \tag{7.35}$$

and during imbibition it obeys

$$k_{rnw} \sim (S_{nw} - S_{rnw})^{\mu/(1+\beta_p)}. \tag{7.36}$$

Similar results are derived for the dispersion coefficients. Near the RS, if holdup dispersion is the dominant mechanism (Chapter 6), then (de Gennes 1983) $D_L \sim (p - p_c)^{-2\nu_p+\mu-\beta_p}$. Therefore, in drainage

$$D_L \sim (S_{nw})^{(\mu-\beta_p-2\nu_p)/\beta_p}, \tag{7.37}$$

and in imbibition

$$D_L \sim (S_{nw} - S_{rnw})^{(\mu - \beta_p - 2\nu_p)/(1 + \beta_p)}. \tag{7.38}$$

These results agree with both simulations and experimental data. For example, (7.37) and (7.38) imply that $D_L \sim S_{nw}^{-0.41}$ and $D_L \sim (S_{nw} - S_{rnw})^{-0.12}$ for drainage and imbibition, respectively. These are in agreement with the experimental data of Delshad *et al.* (1985), which indicate weak divergence of D_L near the RSs.

7.12 Conclusions

Percolation theory has given us a much deeper understanding of two-phase flows in porous media. Not all aspects of two-phase flows in porous media can be explained by percolation. It is also true that the classical random percolation may have to be modified in several significant ways in order to become a quantitative tool for modeling two-phase flows in porous media. But the fundamental concept of percolation – the macroscopic connectivity of a fluid phase – has led us to a much better appreciation of mechanisms of flow phenomena in porous rocks.

References

Anderson, W.G., 1986, *J. Pet. Tech.* **38**, 1246.
Anderson, W.G., 1987, *J. Pet. Tech.* **39**, 1283.
Birovljev, A.L., Furuberg, L., Feder, J., Jϕssang, T., Malϕy, K.J., and Aharony, A., 1991, *Phys. Rev. Lett.* **67**, 584.
Blunt, M., King, M.J., and Scher, H., 1992, *Phys. Rev. A* **46**, 7680.
Buldyrev, S.V., Barabási, A.-L., Caserta, F., Havlin, S., Stanley, H.E., and Vicsek, T., 1992, *Phys. Rev. A* **45**, R8313.
Chandler, R., Koplik, J., Lerman, K., and Willemsen, J., 1982, *J. Fluid Mech.* **119**, 249.
Cieplak, M. and Robbins, M.O., 1990, *Phys. Rev. B* **41**, 2042.
de Gennes, P.G., 1983, *J. Fluid Mech.* **136**, 189.
Delshad, M., MacAllister, D.J., Pope, G.A., and Rouse, B.A., 1985, *Soc. Pet. Eng. J.* **25**, 476.
Dias, M.M. and Wilkinson, D., 1986, *J. Phys. A* **19**, 3131.
Family, F. and Vicsek, T., 1985, *J. Phys. A* **18**, L75.
Fatt, I., 1956, *Trans. AIME* **207**, 144.
Furuberg, L., Feder, J., Aharony, A., and Tϕssang, T., 1988, *Phys. Rev. Lett.* **61**, 2117.
Heiba, A.A., Davis, H.T., and Scriven, L.E., 1983, *Society of Petroleum Engineers Paper 12172*, San Francisco, California.
Heiba, A.A., Sahimi, M., Scriven, L.E., and Davis H.T., 1982, *Society of Petroleum Engineers Paper 11015*, New Orleans, Louisiana.
Heiba A.A., Sahimi, M., Scriven, L.E., and Davis, H. T., 1992, *SPE Reservoir Engineering* **7**, 123.

Hentschel, H.G.E. and Family, F., 1991, *Phys. Rev. Lett.* **66**, 1982.
Horváth, V.K., Family, F., and Vicsek, T., 1990, *Phys. Rev. Lett.* **65**, 1388.
Horváth, V.K., Family, F., and Vicsek, T., 1991a, *J. Phys. A* **24**, L25.
Horváth, V.K., Family, F., and Vicsek, T., 1991b, *Phys. Rev. Lett.* **67**, 3207.
Kardar, M., Parisi, G., and Zhang, Y.-C., 1986, *Phys. Rev. Lett.* **56**, 889.
Kim, J.M. and Kosterlitz, J.M., 1989, *Phys. Rev. Lett.* **62**, 2289.
Kinzel, W., 1983, *Annl. Israel Phys. Soc.* **5**, 425.
Koplik, J. and Levine, H., 1985, *Phys. Rev. B* **32**, 280.
Larson, R.G., Scriven, L.E., and Davis, H.T., 1977, *Nature* **268**, 409.
Larson, R.G., Scriven, L.E., and Davis, H.T., 1981a, *Chem. Eng. Sci.* **36**, 57.
Larson, R.G., Davis, H.T., and Scriven, L.E., 1981b, *Chem. Eng. Sci.* **36**, 75.
Lenormand, R. and Bories, S., 1980, *C. R. Acad. Sci. Paris* **B291**, 279.
Lenormand, R. and Zarcone, C., 1985a, *Phys. Rev. Lett.* **54**, 2226.
Lenormand, R. and Zarcone, C., 1985b, *Physicochem. Hydro.* **6**, 497.
Marle, C.M., 1981, *Multiphase Flow in Porous Media*, Houston, Gulf Publishing.
Melrose, J.C. and Brandner, C.F., 1974, *Can. J. Pet. Tech.* **13**, 54.
Morrow, N.R. and McCaffery, F.G., 1978, in *Wetting, Spreading, and Adhesion*, edited by G.F. Padday, New York, Academic, p. 289.
Payatakes, A.C. and Dias, M.M., 1984, *Rev. Chem. Eng.* **2**, 85.
Richardson, J.G., Kerver, J.G., Hafford, J.A., and Osoba, J., 1952, *Trans. AIME* **195**, 187.
Roux, J.-N. and Wilkinson, D., 1988, *Phys. Rev. A* **37**, 3921.
Roux, S. and Guyon, E., 1989, *J. Phys. A* **22**, 3693.
Rubio, M.A., Edwards, C.A., Dougherty, A., and Gollub, J.P., 1989, *Phys. Rev. Lett.* **63**, 1685.
Sahimi, M., 1988, *Chem. Eng. Commun.* **64**, 179.
Sahimi, M., 1993, *Rev. Mod. Phys.*
Sahimi, M. and Imdakm, A.O., 1988, *J. Phys. A* **21**, 3833.
Sahimi, M. and Tsotsis, T.T., 1985, *J. Catal.* **96**, 552.
Sapoval, B., Rosso, M., and Gouyet, J.F., 1985, *J. Physique Lett.* **46**, L149.
Shaw, T.M., 1987, *Phys. Rev. Lett.* **59**, 1671.
Tang, L.-H. and Leschhorn, H., 1992, *Phys. Rev. A* **45**, R8309.
Thompson, A.H., Katz, A.J., and Rashke, R.A., 1987, *Phys. Rev. Lett.* **58**, 29.
Wilkinson, D., 1984, *Phys. Rev. A* **30**, 520.
Wilkinson, D., 1986, *Phys. Rev. A* **34**, 1380.
Wilkinson, D. and Barsony, M., 1984, *J. Phys. A* **17**, L129.
Wilkinson, D. and Willemsen, J., 1983, *J. Phys. A* **16**, 3365.
Zhang, Y.-C., 1990, *J. Physique* **51**, 2113.

8

Transport, reaction, and deposition in evolving porous media

8.0 Introduction

Earlier chapters discuss flow and transport in porous media. Those phenomena are passive in the sense that they do not cause any changes in the morphology of porous media. In the present chapter we discuss the application of percolation to several phenomena in porous media that cause dynamical changes in the morphology of the pore space. These are, (1) noncatalytic gas–solid reactions with fragmentation; (2) noncatalytic gas–solid reactions with solid products; (3) catalyst deactivation; and (4) flow and deposition of colloidal particles and stable emulsion droplets in porous media.

8.1 Noncatalytic gas–solid reactions with fragmentation

We discuss this phenomenon by giving a specific example. Gasification of a single char particle in CO_2

$$C(s) + CO_2(g) \rightarrow 2CO(g), \tag{8.1}$$

the so-called Boudouard reaction, is an important phenomenon that has been studied for many years. The char particles are porous and contain a large number of small pores. Two important regimes to be considered are the *kinetic regime*, in which the concentration of CO_2 is the same everywhere and diffusion does not play any role, and the *diffusion-limited regime* in which the reactants have to diffuse into the pore space before they react. We first consider the kinetic regime.

8.1.1 Noncatalytic gas–solid reactions in the kinetic regime

Various studies indicate that a Langmuir–Hinshelwood kinetic expression can correlate the experimental data on reaction (8.1), and that the expression reduces to a first order reaction in CO_2, if its partial pressure is low. Therefore, the reaction rate $\hat{R}$ per unit accessible area is written as

$$\hat{R} = k_0 \exp(-E/R_g T), \tag{8.2}$$

where k_0 is the preexponential factor, E is the activation energy, R_g is the gas constant and T is the temperature of the system. Because in the kinetic regime the concentration of CO_2 is the same everywhere, the knowledge of the rate of consumption of carbon is enough for describing the gasification. A simple mass balance on a volume element of C yields

$$\frac{d\phi}{dt} = \frac{M}{\rho} \hat{R} S^A, \tag{8.3}$$

where ϕ is the porosity of the carbon particle, M and ρ are its molecular weight and density, respectively, and S^A is the accessible surface area of the particle. We have assumed that there are no impurities in the carbon particle. It should be clear that CO_2 can only reach that part of the carbon particle that is accessible from the outside, and therefore the accessibility that we define here is exactly in the same sense as in percolation defined in Chapter 2. This makes it clear that coal gasification in the kinetic regime is a percolation phenomenon. Moreover, as (8.3) indicates, to study this problem we only need to keep track of the evolution of the perimeter of the external surface of the particle, i.e., a purely geometrical problem.

An important consequence of gasification is the phenomenon of fragmentation, which can occur both in the kinetic regime and in the diffusion-limited regime, when diffusion is not very fast. If the reaction and consumption of the carbon particle continues, its porosity increases and at some well-defined value ϕ_f the particle fragments into several pieces. In some cases, e.g., fragmentation during devolatilization, we might have a pressure-induced fracture and fragmentation of the particle. But if the consumption rate of the particle is low enough, as is the case in most practical situations, fracture-induced fragmentation can be safely neglected. Fragmentation in the diffusion-limited regime is usually called *perimeter fragmentation*, since it occurs mainly on the external surface or perimeter of the particle.

Fragmentation of coal particles is believed to be responsible for decreased burnout times, enhanced production of fly ash, weight loss in coal combustors, and increased emission of submicrometer unburnt carbon and NO_x from pulverized coal combustion systems. This phenomenon was studied experimentally by Sundback *et al.* (1984), Kerstein and Niksa (1984), and Sadakata *et al.* (1984). Particle fragmentation can be inferred from image analysis of char particles following coal devolatilization (Sundback *et al.* 1984), or from size measurements of char particles retrieved from fluidized beds and char particle breakup (Kerstein and Niksa 1984). In the early studies of gasification, fragmentation was either ignored completely, or was dealt with in an empirical way by treating ϕ_f as an adjustable parameter of the model without any regard for particle morphology. For example, based on experimental observations Gavalas (1981) assumed that, $\phi_f \simeq 0.8$. This value of ϕ_f implies that the critical volume fraction of the solid matrix for fragmentation is about 0.2, close to 0.17, the critical volume fraction for percolation of continua discussed in Chapter 2.

8.1.2 Noncatalytic gas–solid reactions in the diffusion-limited regime

In this regime, transport of the reactants into the pore space both by ordinary and pressure diffusion controls the overall mass transfer of the reactants and the gasification process. We introduce two effective transport coefficients L_{ij}^D and L_{ij}^P, which describe the contributions of diffusion and permeability of the pore structure of the particle, where ij refers to transport of component i in the mixture of i and j. These coefficients are given by (Jackson 1977)

$$L_{ij}^D = \frac{1}{1/D_{Ki} + (1 + \delta_{ij}X_i)/D_{ij}(P)}, \tag{8.4}$$

$$L_{ij}^P = \frac{1 + D_{Ki}/D_{ij}(P)}{1/D_{Ki} + (1 + \delta_{ij}X_i)/D_{ij}(P)} + \frac{k_p P}{\eta_m}. \tag{8.5}$$

Here D_{Ki} and D_{ij} are, respectively, the effective Knudsen and pressure diffusivities at pressure P, X_i the mole fraction of i, $\delta_{ij} = (M_i/M_j)^{1/2} - 1$ essentially represents the ratio of the molecular weights of i and j, k_p is the permeability of a single pore, and η_m is the effective viscosity of the mixture. The evolution of the system is described by a diffusion reaction equation (8.6) that tells us how the reactant is transported within the pore space, and how the particle is consumed. Moreover, we need a mass balance at the external surface of the particle to express the fact that the diffusive flux there is equal to the rate of mass transfer (8.9). Thus, for a spherical particle the concentration C_i of component i at time t is governed by

$$\frac{\partial C_i}{\partial t} = \frac{1}{R_g T}\frac{1}{r^2}\frac{\partial}{\partial r}\left(r^2 L_{ij}^D P\frac{\partial X_i}{\partial r} + r^2 L_{ij}^P X_i \frac{\partial P}{\partial r}\right) + s_i \hat{R} S^A, \tag{8.6}$$

$$C_i(r, 0) = 0, \tag{8.7}$$

$$\frac{\partial C_i}{\partial r}(0, t) = 0, \tag{8.8}$$

$$\frac{1}{R_g T}\left(L_{ij}^D P \frac{\partial X_i}{\partial r} + L_{ij}^P X_i \frac{\partial P}{\partial r}\right) = K(C_{ib} - C_i^f), \tag{8.9}$$

where s_i is the stoichiometric coefficient of i in the reaction, K is the mass transfer coefficient, C_{ib} is the concentration of i in the bulk, and C_i^f is the concentration of i at a radial position r_f where fragmentation occurs. If the process is highly nonisothermal, then the transport coefficients depend on temperature, and we have to add the energy equation to the above model. Because the pore space evolves during gasification and the porosity increases, the transport coefficients also vary with ϕ. Thus, we have to estimate them as a function of ϕ, a problem discussed in Chapter 5.

We now write down a mass balance for a particle of radius R

$$\frac{\partial \phi}{\partial t} = \frac{M}{\rho} \hat{R} S^A + \frac{\partial \phi}{\partial r}\frac{\partial R}{\partial t}, \tag{8.10}$$

with the initial condition that, $\phi(r, 0) = \phi_0$, where ϕ_0 is the initial porosity of the system. The second term of the right side of (8.10) is due to the fact that the radius of the particle shrinks with time as a result of the reaction on the external surface of the particle and perimeter fragmentation. The contribution of the reaction can easily be accounted for by writing a mass balance on the external surface of the particle which yields

$$\frac{dr_f}{dt} = -\frac{M}{\rho}\hat{R}, \tag{8.11}$$

with $r_f(0) = r_0$. If perimeter fragmentation starts at a time t_f, then

$$\frac{dr_f}{dt} = -\frac{M\hat{R}\partial\phi/\partial r}{\rho S^A}, \tag{8.12}$$

with, $r_f(t_{f-}) = r_f(t_{f+})$. Obviously, for $t < t_f$ (8.12) should not be used because fragmentation has not begun yet.

8.2 Percolation models of noncatalytic gas–solid reactions and fragmentation

How do we model this phenomenon? Theoretical investigation of this process goes back to Petersen (1957) who modeled the pore structure as an idealized network of randomly intercepting cylindrical pores. Many more models have been developed since Petersen's work which have been completely discussed by Sahimi *et al.* (1990). Here we are only interested in percolation models of gasification. There have been two types of such models for reactive porous media and, in particular, gasification. The first type is what we call *hybrid continuum models*, while the second type are the usual *network* models used in percolation modeling of many phenomena.

8.2.1 Hybrid continuum models

These models are based on solving (8.6)–(8.9) and calculating the consumption rate of the coal particle. At each time step, (8.6)–(8.9) are solved numerically. Since the porosity of the system increases with time, the effective morphological properties (e.g., accessible surface area) and transport coefficients of the system also vary with porosity and time. This variation with porosity is of course a percolation effect and is included in the model by assuming a certain model of the pore space and calculating all the percolation quantities that are needed for use in (8.6)–(8.9). As such, these models are a hybrid of the classical equations of transport and reaction, and percolation methods of estimating the effective transport properties of a disordered medium discussed in Chapter 5. Mohanty *et al.* (1982) were the first to develop a percolation model of reactive porous media. They represented the porous medium by a cubic tessellation in which a randomly selected fraction p of the cubic polyhedra represented the pores. The increase in the porosity as a result of gasification was modeled by simply increasing p (i.e., gasification was modeled as a random percolation process). This problem is similar to immiscible displacement of a fluid by an invading fluid and invasion percolation, discussed in Chapter 7, in which there is a well-defined interface between the displaced and displacing fluids. There is a *reaction front* between the reactants and the solid matrix. As a result, the increase in the porosity should be modeled by consuming the solid polyhedra *adjacent to the percolating pore cluster or the reaction front*. As such, the model of Mohanty *et al.* is not completely satisfactory. Reyes and Jensen (1986) improved this model by representing the pore space as a Bethe network of cylindrical pores with distributed radii, to take into account the effect of the pore size distribution of the particle, but they also modeled the porosity increase by random percolation. For the kinetic regime, they obtained good agreement between the predicted reaction rates, at various values of conversion (i.e., the consumed fraction of the particle), and experimental data; see Fig. 8.1. Shah and Ottino (1987a) also modeled gasification by solving (8.6) and (8.9), but representing the pore space as a cubic network. But unlike the earlier models, porosity was increased in a physical way by monitoring the evolution of the morphology and the reaction front that was constantly moved to the edge of the remaining matrix.

The hybrid continuum models are subject to a fundamental criticism. (8.6)–(8.9) may not be valid near the fragmentation point, which is a percolation threshold. As ϕ_f is approached (i.e., as the matrix is consumed and ϕ is increased), the percolation correlation length ξ_p, pertaining to the

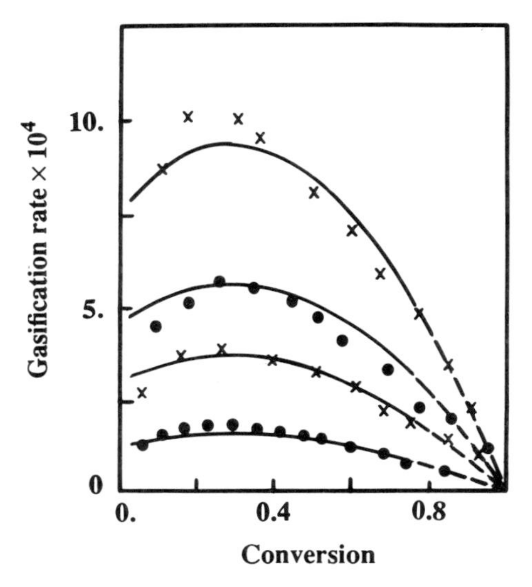

Figure 8.1 *Comparison of percolation predictions of char-air gasification rates (in* s^{-1}*) with the experimental data (symbols). The pore space was modeled by a Bethe lattice of coordination number* $Z = 7$*. The curves are, from top, for temperatures 753K, 728K, 703K, and 673K (after Reyes and Jensen 1986).*

solid matrix, also increases without bound. Thus, over length scales $L < \xi_p$ the solid matrix has a fractal structure, in which case continuum equations such as (8.6) and (8.9) are not valid.

8.2.2 Network models

These models were developed by Kerstein and Bug (1986), Kerstein and Edwards (1987), and Sahimi and Tsotsis (1987, 1988). The details of their models are not the same, but the main ideas are identical. We start with a network in which a fraction p of the bonds represents the pores; the rest represent the solid matrix. We identify the solid bonds on the external perimeter of the pore space. They are then consumed and redesignated as pores, the process time is increased by one unit, the perimeter solid bonds in the new configuration of the network are identified, and so on. If the consumption of the solid bonds is continued, then at a porosity ϕ_f the solid particle disintegrates into finite fragments (clusters) with a wide variety of shapes, sizes, and masses. The precise value of ϕ_f depends on the microscopic details of the solid matrix and its chemical composition, just as p_c depends on the coordination number Z of the network. For example, if a portion of the matrix were nonreactive, fragmentation would occur at a lower ϕ_f. The presence of such impurities can easily be accounted for by designating a fraction of the bonds as nonreactive. Figure 8.2 shows the dependence of the reaction rate, i.e., the fraction of consumed solid bonds

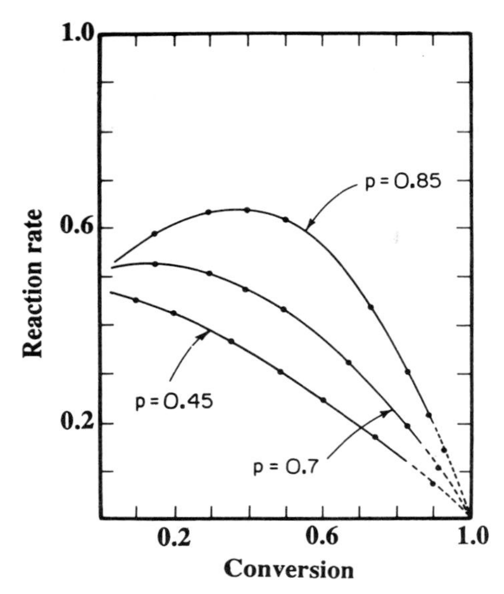

Figure 8.2 *The dependence of the reaction rate on the initial fraction p of solid bonds in a percolation model of gasification, compare this with Fig. 8.1 (after Sahimi 1991).*

per unit time, on total conversion in a three-dimensional network with $Z = 14$. For small values of p, we observe a maximum in the reaction rate, which is observed frequently in the experiments. This maximum is caused by two competing geometrical factors. At low values of conversion, the consumption of the matrix *increases* the length of the perimeter of the solid matrix (cluster of solid bonds). This causes an increase in the reaction rate proportional to the perimeter length. At high conversions, the length of the perimeter *decreases* with consumption. This causes the reaction rate to decrease monotonically. Kerstein and Bug (1986) showed that in this model the solid matrix is consumed in a time t_c where

$$t_c \sim \ln(\varepsilon^{-1}), \tag{8.13}$$

and $\varepsilon = \phi_f - \phi$, so that we have a simple way of estimating the burning time as a function of the porosity.

How do we account for the effect of diffusion? Kerstein and Edwards (1987) assigned to each solid bond a random burning time selected from an exponential distribution, $f(t) = \lambda \exp(-\lambda t)$, and related λ to the mass m of the solid matrix via, $\lambda = am^{-b}$, where b is an adjustable parameter. The idea is that different portions of the particle have different burning times, because the diffusing reactants reach them later. After fragmentation, a distinct value of λ is used for each fragment. Although the model is somewhat ad hoc, it does appear to provide reasonable fit to some experimental data.

In the model of Sahimi and Tsotsis (1987), diffusion is represented by a random walk of the reactants that are injected into the system at the external surface of the network. The random walk is executed on the bonds that represent the pores, and the probability of selecting any pore is proportional to its conductance. When the reactant hits a solid bond it reacts with it with probability $p_r = \hat{R}/k_0$, i.e., the normalized reactivity of the matrix. Thus, the effect of pore size distribution, temperature, and other influencing factors are taken into account. If reaction does take place, the reactive molecule disappears and a fraction of the solid bond is consumed. Since we use finite networks to represent char particles of micrometer size, each bond represents a macroscopic mass unit. Thus, each solid bond has to be hit by the molecule, in a reactive collision, n_c times (n_c is a parameter), before it is totally consumed. If $p_r \simeq 1$, then only the most exposed part of the matrix is consumed at the initial stages of the process. The net effect is the removal of all irregularities of the external surface of the matrix, making its shape smooth and regular. Thus, at most, perimeter fragmentation can occur. On the other hand, if p_r is small, the reactive molecules penetrate deep into the pore space and consume the weak points of the matrix. As a result, matrix fragmentation occurs and many fragments with a wide variety of shapes and sizes appear. Thus, the limits $p_r = 1$ and $p_r \simeq 0$ represent, respectively, the diffusion-limited and the kinetic regimes. Varying p_r be-

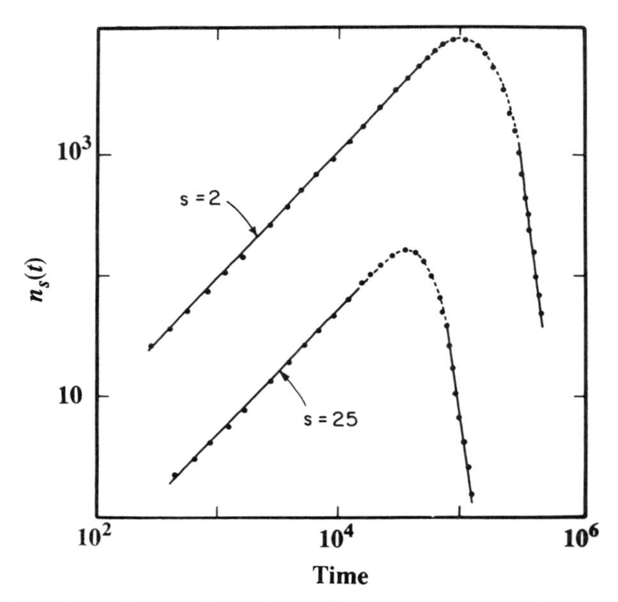

Figure 8.3 *Time-dependence of number of fragments $n_s(t)$ of size s. The reactive porous medium was modeled by the sample-spanning cluster on a simple cubic lattice at its site percolation threshold (after Sahimi 1991).*

tween zero and one enables us to study a variety of situations in which the relative importance of kinetic and transport effects can be varied.

To describe the evolution of fragment size distribution, a dynamic scaling of the following form is assumed (Sahimi and Tsotsis 1987, Sahimi 1991)

$$n_s \sim t^w s^{-\tau} f(s/t^z), \tag{8.14}$$

where n_s is the number of fragments (per network bond) of s bonds, w, z, and τ are dynamical exponents, and $f(x)$ is a scaling function. These exponents are universal and do not depend on the microscopic details of the system, e.g., the pore size distribution of the particle. The exponent w is positive before a characteristic time t_m is reached, at which the number of fragments reaches a maximum, and is negative for $t > t_m$. The quantity t^w reflects the time evolution of the number of fragments, and distinguishes (8.14) from the percolation equation (2.14). A similar equation has also been used in polymer chain degradation, and even porous rock fragmentation (Englman *et al.* 1984). Thus, (8.14) may be a general property of *all* fragmentation processes. Figure 8.3 shows the variations with time of $n_s(t)$, calculated on a cubic network. It indicates that $w = 1$. Cai *et al.* (1991) and Gyune and Edwards (1992) developed a scaling theory and a linear rate equation for fragmentation with mass loss and derived (8.14).

8.3 Noncatalytic gas–solid reactions with solid products

This class of phenomena is encountered in several important chemical processes. For example, limestone is used to control SO_2 emissions from fluidized-bed combustors. Metal oxides are used in coal gas desulphurization processes. In both cases solid products are formed by noncatalytic gas–solid reactions. When the stoichiometric volume ratio of the system, i.e., the ratio of the molar volume of the solid product and stoichiometrically equivalent volume of the solid reactant, exceeds one, the solid product occupies more space than the original solid, the porosity of the porous medium decreases as the reaction continues and pore plugging occurs, implying that this is a percolation phenomenon. Similar to gasification with fragmentation, this problem has also been investigated by many authors, using a variety of ideas and methods. These models have been reviewed by Sahimi *et al.* (1990). As usual, we are only interested in the percolation models of this phenomena, which we now discuss.

During any noncatalytic gas–solid reaction with solid products in a porous structure (and also in catalyst deactivation discussed in the next section), all pores of the porous medium can be divided into three distinct groups (Sahimi and Tsotsis 1985): (1) Pores that are completely plugged; (2) pores that are partially plugged (with reduced radii), and are still accessible to the reactant; and (3) pores that are not completely plugged but are surrounded by plugged pores and cannot be reached by the reactants. The first difference between noncatalytic gas–solid reactions and random percolation is that some of the open pores are *trapped* by the plugged pores in noncatalytic gas–solid reactions. Therefore, this is a percolation phenomenon with trapping, discussed in Chapter 7.

We now consider a specific example and develop a percolation model for it. We consider the sulphation of calcined limestone particles, which is given by the reaction

$$CaO(s) + SO_2(g) + \frac{1}{2}O_2(g) \rightarrow CaSO_4(s). \tag{8.15}$$

Experimental data show that during this reaction the solid volume increases by a factor of about 3.1, and that in excess O_2, the reaction is first order with respect to the partial pressure of SO_2. Similar to gasification, we can consider both the kinetic and diffusion-limited regimes. But in practice the kinetic regime is hardly ever realized, therefore we develop the model for the diffusion-limited case. A percolation model for the kinetic case is given by Yortsos and Sharma (1986) and Yu and Sotirchos (1987). For the most recent study see Sotirchos and Zarkanitis (1993). In the diffusion-limited regime, the governing equations for a spherical porous medium are given by (Reyes and Jensen 1987)

$$\frac{\partial C_1}{\partial t} = \frac{1}{r^2}\frac{\partial}{\partial r}\left(r^2 D_1 \frac{\partial C_1}{\partial r}\right) - \hat{R} C_2 S^A, \tag{8.16}$$

$$\frac{\partial C_1}{\partial r}(0, t) = 0, \tag{8.17}$$

$$D_1 \frac{\partial C_1}{\partial r} = K(C_{1b} - C_{R_0}), \tag{8.18}$$

where D_1 is the effective diffusivity of SO_2–O_2, C_1 and C_2 are the SO_2 concentrations at the gas–solid and solid–solid interfaces, respectively, and the rest of the notation is as before. Equations (8.16)–(8.18) are similar to (8.6)–(8.9), except that for the present case the contribution of pressure diffusion is negligible. We need an equation for C_2 to express its variations within the solid layer that is building up in every pore. Since transport within the layer is by diffusion, we must have

$$\frac{\partial}{\partial \zeta}\left(\zeta^{\delta-1} D_l \frac{\partial C_2}{\partial \zeta}\right) = 0, \tag{8.19}$$

$$D_l \frac{\partial C_2}{\partial \zeta}(r_{ss}) = \hat{R} C_2(r_{ss}), \tag{8.20}$$

$$C_2(r_{ss}) = C_1, \tag{8.21}$$

where D_l is the diffusivity within the layer, ζ is measured in the direction perpendicular to the reacting surface, r_{ss} is the radius of the solid–solid interface, and δ is a geometrical factor such that $\delta = 1, 2, 3$ for slabs, cylinders, and spheres, respectively. Experimental data for D_l are used in (8.19) and (8.20), because its theoretical estimation is difficult.

We also need an equation for the consumption rate of CaO. Suppose that ϕ_s is the volume fraction of CaO in the particle. Then, a simple mass balance gives

$$\frac{d\phi_s}{dt} = -\frac{M_s}{\rho_s}\hat{R}C_2(r_{ss}, t)S^A, \tag{8.22}$$

where M_s and ρ_s are the molecular weight and density of the solid, respectively, and $\phi_s(t = 0) = \phi_0$. Because of the formation of solid products, the pore radius $R(r, t)$ continuously shrinks, and since volume is conserved, $R(r, t)$ is related to the initial pore radius R_0 and r_{ss} by the following equation

$$R(r, t) = e_r R_0 + (1 - e_r) r_{ss}(r, t), \tag{8.23}$$

where e_r is the expansion coefficient (about 3.1). The last thing to do is to write down an evolution equation for the pore size distribution $f(r, t)$ of the porous medium. Only accessible pores are of interest to us, and therefore

$$\frac{\partial[\phi^A f(r, t)]}{\partial t} + \frac{\partial}{\partial r}\left[\phi^A f(r, t)\frac{dR}{dt}\right] = \frac{d\phi^A}{dt} f(r, t), \tag{8.24}$$

where ϕ^A is the accessible porosity of the system, $f(r, 0) = f_0(r)$, and $f(r, \infty) = 0$, where $f_0(r)$ is the initial pore size distribution of the pore space. The second boundary condition arises because after a long time all pores will be plugged.

The evolution with time or position of all quantities of interest can now be calculated. Percolation enters this model in two different ways. One is through D_1, the effective diffusivity of the SO_2–O_2 pair. Since the pores are plugging, D_1 has to be calculated for a percolation network whose porosity is shrinking, a problem discussed in Chapter 5. Percolation also enters the model through ϕ^A, the accessible porosity (accessible fraction of the pores), which also decreases as the system evolves. To use the model, assume a network model of pore space and estimate its ϕ^A and D_1 as the pore space evolves (as ϕ decreases), substitute them into the above equations, and solve for the properties of interest.

This is a hybrid continuum model, and therefore it is also subject to the same criticism as that for gasification. But its predictions are reasonably

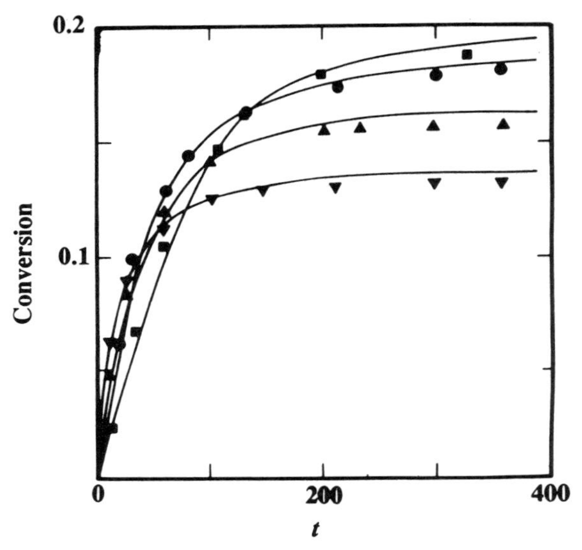

Figure 8.4 *Percolation predictions of CaO conversion versus time t (in min) and their comparison with the experimental data (symbols). The results are, from top, at temperatures 1023K, 1113K, 1163K, and 1213K (after Reyes and Jensen 1987).*

accurate. For example, Fig. 8.4, adopted from Reyes and Jensen (1987), compares the predicted conversion of CaO as a function of time with the experimental data. The pore space model was assumed to be a Bethe network with $Z = 6$. For short times, the agreement is excellent, but at longer times the predictions and the experimental data show some difference. Other variations of this basic model, as well as other approaches to this phenomenon are reviewed by Sahimi *et al.* (1990).

8.4 Catalyst deactivation

Another phenomenon to which the concepts of percolation theory have been applied is deactivation of porous catalysts. This phenomenon is typically caused by a chemical species which adsorbs on and poisons the catalyst's surface and its active sites, where catalytic reactions occur, and frequently blocks its pores. We often find that reactants, products, and reaction intermediates, as well as various reactant stream impurities, also serve as poisons and/or poison precursors. As a result, the morphology of the catalyst evolves with time, and its pore volume decreases. After some time, no sample-spanning cluster of open pores exists, and the catalyst deactivates and loses its effectiveness, which means that catalyst deactivation is a percolation process. Because of its industrial significance, numerous theoretical and experimental investigations have been devoted to catalyst deactivation. A general review of the subject is given by Hughes (1984). As usual, we are interested only in the percolation models of catalyst deactivation.

Early studies of catalyst deactivation either used a bundle of parallel capillary tubes to represent the pore space of the catalyst, or employed continuum equations of transport and reaction in which the transport coefficients and other important parameters were inputs to the model. But these models were inadequate because they could not take into account the important effect of the interconnectivity of the catalyst pore space and its time evolution during deactivation. Improvements to these models began in the late 1970s. An interesting class of catalyst deactivation models was developed by Froment and coworkers (see Nam and Froment 1987, Beeckman *et al.* 1987, for references to their earlier papers), in which probabilistic arguments were used to derive expressions for the activity or reaction rate of the catalyst. To represent the pore space, they used a variety of models, such as a single pore, the bundle of parallel pores, and a Bethe lattice with $Z = 3$. Using this lattice, Froment and coworkers were the first to recognize that, in agreement with the experimental observations, the catalytic activity of a catalyst undergoing deactivation vanishes much earlier than predicted by the bundle or parallel tube model. This is caused by the interconnectivity of the pores, the main idea in percolation, but they did not recognize it. Catalyst deactivation was first identified as a percolation phenomenon by Sahimi and Tsotsis (1985). To discuss percolation models of catalyst deactivation, we again distinguish between the kinetic and diffusion-limited regimes.

8.4.1 Percolation model of catalyst deactivation in the kinetic regime

For concreteness, we discuss the model that was developed by Sahimi and Tsotsis (1985). A porous catalyst, in an isothermal reactor, with the catalytically active material uniformly distributed in its pores (at an initial concentration C_0) is reacting while simultaneously undergoing slow deactivation. The overall reaction rate $\hat{R}$ in a single pore of radius R and length l_p of the catalyst is given by

$$\hat{R} = 2\pi R l_p \hat{R}_0 (C_0 - C_s)^m \tag{8.25}$$

where $C_0 - C_s$ is the active site concentration at time t, $\hat{R}_0$ is the reaction rate per unit area, and m is the order of reaction. A parallel deactivation reaction process results in the deposition of a contaminant (a deposit with an average volume b per unit weight of sites), which poisons active sites and simultaneously blocks part of the pore volume. The rate of change in the concentration of poisoned sites is given by

$$\frac{dC_s}{dt} = \hat{R}_d(C_0 - C_s)^n \tag{8.26}$$

where $\hat{R}_d$ is the deactivation reaction rate per unit area, and n is the reaction order. The chemistry of the reactions is embedded in $\hat{R}_0$ and $\hat{R}_d$. To a partially plugged pore with the initial radius R_0 we assign an effective radius R_e, the radius of a cylindrical pore with the same open volume and length, and is given by

$$R_e = R_0\left[1 - \frac{\alpha}{R_0} g(\theta)\right]^{1/2} \tag{8.27}$$

where

$$g(\theta) = \begin{cases} 1 - [1 + (n-1)\theta]^x & n < 1, \quad \theta < x, \\ 1 - \exp(-\theta) & n = 1, \\ 1 - [1 + (n-1)\theta]^x & n > 1 \end{cases}$$

with $x = (1 - n)^{-1}$, and

$$\alpha = 2C_0 b, \tag{8.28}$$

$$\theta = \hat{R}_d C_0^{n-1}. \tag{8.29}$$

where θ is the dimensionless time, and α is the effective size of the deposit. If $\alpha < R_0$, $R_e > 0$ at all times, while for $\alpha \geqslant R_0$, $R_e = 0$ after a finite time θ_p. This establishes a one-to-one correspondence between R_0 and its plugging time θ_p. A pore of initial radius R_0 is plugged at time θ_p as long as $R_0 < \alpha g(\theta)$. As the deactivation proceeds, larger and larger isolated islands of partially plugged pores appear that are surrounded by the completely plugged pores. Such pores can no longer contribute to the catalytic activity of the catalyst. At or below a critical volume fraction of the plugged pores, we intuitively expect no sample-spanning cluster of open pores to exist. Realistically of course, the assumption of kinetic control may not be valid near this point, because after some pores are plugged, the reactants have to diffuse in the pore space to reach the active sites of the catalyst. Thus, the model is oversimplified, but it is useful because it clearly demonstrates the role of percolation during catalyst deactivation. Similar to noncatalytic reactions with solid products, catalyst deactivation is a percolation phenomenon with *trapping*. In both phenomena, the effect of trapping is only significant if the pore size distribution of the pore space is broad enough to allow for a wide variety of pore sizes in the catalyst. Otherwise, the effect of trapping is small and the deactivation process is essentially the same as the random bond percolation discussed in Chapter 2.

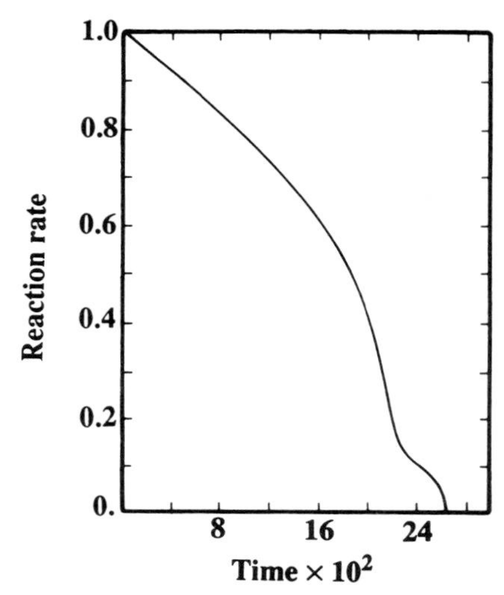

Figure 8.5 *Reaction rate versus dimensionless time during a kinetically controlled deactivation process. The reaction rate has been normalized by its value when no pore has been plugged yet. The catalyst was modeled by a simple cubic network with a log-normal pore size distribution (after Sahimi and Tsotsis 1985).*

How do we simulate this process? We start with a pore network with distributed pore sizes, estimate the effective radii of the pores at any given time (which are decreasing with time) using (8.27), and identify the completely plugged pores, the partially plugged but accessible pores, and the trapped pores. In the subsequent steps of simulations, the trapped pores are ignored, since they cannot be reached by the reactant. The simulations continue until a sample-spanning cluster of open pores no longer exists. Normally, the reaction rate of the catalyst is proportional to its *accessible* surface area. Figure 8.5 presents the time variations of the reaction rate of a catalyst calculated with a simple cubic network and a log-normal pore size distribution. It is clear that the reaction rate vanishes at a finite θ_p. The knee of the curve is typical of deactivation processes in the kinetic regime, which has been observed in experiments.

8.5 Catalyst deactivation in the diffusion-limited regime

Catalyst deactivation is usually accompanied by diffusional limitations. Similar to gasification, percolation models of catalyst deactivation in this regime can also be divided into two groups: the hybrid continuum models, and the network models.

8.5.1 Hybrid continuum models

In this approach, we use differential equations of diffusion and reaction, similar to those used for gasification and noncatalytic gas–solid reactions with solid products. For concreteness, we discuss and develop a model of catalyst deactivation for hydrodemetallation (HDM). During catalytic hydroprocessing for sulfur and nitrogen removal, nickel and vanadium porphyrins, which are the principal metals in petroleum and other metal-bearing molecules, undergo HDM reactions leading to metal deposition and catalyst poisoning. It has been suggested that the mechanism of HDM involves an intermediate B

$$A \underset{k_2}{\overset{k_1}{\rightleftharpoons}} B \overset{k_3}{\rightarrow} C \tag{8.30}$$

where A represents the metalloporphyrin compound in oil, B the hydrogenated intermediate metallochlorin in oil, C metal deposited on the pore surface, and k_1, k_2, and k_3 are three kinetic constants. The reactions are believed to be first order, and therefore the concentrations C_A and C_B of A and B obey the following diffusion reaction equations

$$\frac{\partial C_A}{\partial t} = \nabla \cdot (D_A \nabla C_A) - k_1 C_A + k_2 C_B, \tag{8.31}$$

$$\frac{\partial C_B}{\partial t} = \nabla \cdot (D_B \nabla C_B) + k_1 C_A - (k_2 + k_3) C_B. \tag{8.32}$$

In spherical coordinates, these equations are similar to (8.6) and (8.16). The main difference between catalyst deactivation and noncatalytic gas–solid reactions with solid products is that, catalytic reactions occur *only* at the active sites of the catalyst, and therefore the deposition of the solid products on the pore surfaces may not be as simple as that in the noncatalytic cases.

During deactivation the deposition mode in any pore can be either *uniform* or *discrete*. In the first case we assume that the thickness of the solid layer formed on the pore surface is the same everywhere. This implies that the fraction of the active sites covered by the deposits increases rapidly from zero to one. Physically, this is possible if the deposit is porous enough to allow the reactants to reach the active sites, or if it contains catalytically active materials itself. In discrete deposition we assume that a fraction of the catalyst's active sites becomes nucleated by the product, after which deposition occurs only on such sites. Therefore, we no longer have a uniform layer of deposits on the pore surface, but a series of discrete lumps. Once we make up our mind about the mode of pore deposition, we can proceed to calculate various properties of the catalyst as deposition takes place. For example, in uniform pore deposition, the pore radius R changes with time according to

$$\frac{\partial R}{\partial t} = -\frac{M_d \hat{R}_d \gamma \theta_c}{\rho_d}, \tag{8.33}$$

with $R(t = 0) = R_0$, where M_d and ρ_d are the molecular weight and density of the deposits, respectively, $\hat{R}_d$ is the rate of deposit production per unit area, γ is the degree of catalytic activity relative to the original surface, and θ_c is the fraction of the active sites covered by the deposits. We have assumed that the pore is cylindrical and that the reaction is first order. Note the similarity between (8.3), (8.11), (8.12), (8.22) and (8.33).

The following steps are taken. We assume a model of the pore space to determine its effective diffusivity during various stages of deactivation. We solve (8.31) and (8.32) numerically. We use (8.33) to keep track of the evolution of the pore sizes, and thus update the structure of the pore network. The numerical solution of (8.31) and (8.32) stops when the pore network reaches its percolation threshold. This approach was developed by Shah and Ottino (1987b), Melkote and Jensen (1989), and Beyne and Froment (1990). It is relatively successful but is subject to the same criticism discussed for gasification and noncatalytic gas–solid reactions with solid products regarding the applicability of the continuum equations as the percolation threshold is approached.

8.5.2 Network models

In this approach we represent the pore space as a network of interconnected pores with distributed pore sizes. If the pores are cylindrical, and if diffusion is mainly in the axial direction, then the pore diffusion and reaction equations are simplified to

$$\frac{\partial C_A}{\partial t} = D_{Ap} \frac{\partial^2 C_A}{\partial x^2} - k_1 C_A + k_2 C_B, \tag{8.34}$$

$$\frac{\partial C_B}{\partial t} = D_{Bp} \frac{\partial^2 C_B}{\partial x^2} + k_1 C_A - (k_2 + k_3) C_B. \tag{8.35}$$

These equations are valid for every pore provided that for every pore $l_p/R >> 1$. Normally, a steady state condition is assumed because the changes in the catalyst pore space are much slower than diffusion and reaction. Note that D_A and D_B are *macroscopic* diffusivities, whereas D_{Ap} and D_{Bp} represent *microscopic* or pore-level diffusivities. Equations (8.34) and (8.35) are then solved with the boundary conditions $C_A(x = 0) = C_{A0}$, $C_A(x = l_p) = C_{Al}$, $C_B(x = 0) = C_{B0}$, and $C_B(x = l_p) = C_{Bl}$, where the boundary concentrations are still unknown. To obtain the concentration profile of C in the entire network, we write down a mass balance for every node of the network

$$\sum_i S_{ij}(J_A)_{ij} = 0, \tag{8.36}$$

$$\sum_i S_{ij}(J_B)_{ij} = 0, \tag{8.37}$$

where S_{ij} is the cross-sectional area of pore ij, and J_A is the diffusive flux of A. Using the analytical solutions of (8.34) and (8.35) the fluxes J_A and J_B are determined and substituted into (8.36) and (8.37). These fluxes contain the nodal concentrations C_{A0}, C_{Al}, C_{B0}, and C_{Bl}. Writing (8.36) and (8.37) for every node of the network results in a set of simultaneous equations for the nodal concentrations. By solving these equations we obtain the concentration profiles of A and B, and thus $C_c = \tau_m k_3 C_B$ in the entire network. The quantity τ_m is the measurement time. Once the concentration profile of C in every pore is known, a pore deposition mode is assumed and the deposit profile in every pore is calculated. Normally, the reactive molecules are relatively large with an effective diameter R_m comparable to those of the pores. If so, the pore diffusivities D_{Ap} and D_{Bp} are a function of $\lambda = R_m/R$, and since metals or coke continuously deposit on the pore surface and reduce R, the pore diffusivities also decrease continuously. In this case, we can use an equation that relates the pore diffusivity to λ; a well-known example is given by (Brenner and Gajdos 1977)

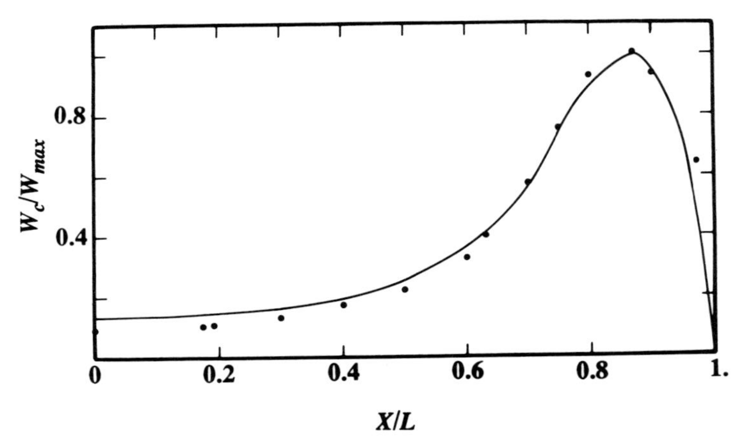

Figure 8.6 *Comparison of percolation predictions of the weight W of the deposit C during an HDM process with the experimental data (circles). W_{max} is the maximum weight of C in the profile, and X is the distance from the center of the catalyst particle. The particle was modeled by an L × L × L cubic network (after Arbabi and Sahimi 1991).*

$$\frac{D_{Ap}}{D_{A\infty}} = 1 + 1.125\lambda\ln\lambda + 0.416\lambda + 2.25\lambda^2\ln\lambda, \qquad (8.38)$$

which is accurate for $\lambda \leqslant 0.4$, where $D_{A\infty}$ is the bulk diffusivity (in an unbounded fluid).

This network model is valid both near and far from the percolation threshold of the catalyst (i.e., the deactivation point), and is free of the limitations of the hybrid continuum models. This model was developed by Arbabi and Sahimi (1991) and can be used for *any* pore structure or pore size distribution, and *any* diffusion reaction mechanism. Figure 8.6 compares the weight profile of C in a catalyst particle, represented by a cubic network of linear size L, with the experimental data for an HDM process. Given the complexities of such phenemona, the agreement is excellent. Arbabi and Sahimi (1991) also showed that their model can be used for studying deactivation with kinetic schemes more complex than (8.30).

8.6 Flow and deposition of colloidal particles and stable emulsions in porous media

So far we have discussed phenomena that change the structure of a porous medium by chemical reactions. We now discuss a class of phenomena that change the morphology of a porous medium not by a chemical reaction but by physical interactions. The physical interactions are between the pore surfaces and the fluid and between the pore surfaces and the fluid contents. This class of phenomena includes deep-bed filtration, migration of fines (small, solid, and electrically charged particles), and flow of stable emulsions in a porous medium. These phenomena occur in many processes of industrial significance such as gel permeation and enhanced recovery of oil. Since all three phenomena are in many ways similar, we only discuss fines migration, and only note its differences with the other two.

Fines migration is caused by the contact of the solid matrix of a porous medium with an incompatible brine solution. If a sandstone core, saturated with a brine solution, is flushed with brine solution and followed by the injection of distilled water, the permeability of the core decreases. The reason is that fines, in the form of clays or other small particles, deposit on the surface of the pores and reduce the porosity, and thus the permeability. This causes severe problems for enhanced recovery of oil as the reduction in the permeability decreases the amount of oil that can be produced.

Stable emulsions, on the other hand, are generated in situ in the porous medium, or sometimes they are injected into it. In miscible displacement of oil by an agent, if the viscosity of the displacing fluid is less than that of oil, the process would be unstable and very inefficient. To make the process more stable and efficient, stable emulsion droplets are used to increase the viscosity of the displacing fluid. But the droplets can deposit on the surface of the pores and reduce the permeability of the pore space. As such, the phenomenon is very similar to fines migration. The main difference between the two is that emulsions only form monolayers on the pore surface, whereas fines can deposit on top of each other. Obviously, in both problems the reduction in the permeability of the pore space is a percolation process.

How do we model this class of phenomena? As with the other three phenomena discussed above, there have been many studies of these processes by a variety of methods and techniques. Most of them have used a continuum approach in which the behavior of the phenomena is approximated by differential equations with no regard for the pore space structure. These studies were reviewed by Tien and Payatakes (1979), and are not of interest to us. More recently, a few models have been developed in which the pore space is represented by a network of interconnected pores, to take into account the effect of the pore space topology and percolation. These models have been reviewed and discussed by Sahimi *et al.* (1990). Here, we discuss one such model.

The percolation model we discuss was developed by Sahimi and Imdakm (1991) and Imdakm and Sahimi (1991). In their model, the porous medium is represented by a network of interconnected cylindrical pores with distributed radii, and the calculations are carried out in several stages. In the first stage, the initial flow field within the network is computed (see Chapters 5 and 6). Particles are assumed to be spherical with distributed radii; they are selected from a particle size distribution, then injected into the network. The exact trajectory of each particle within a pore is then determined by writing a force balance for each particle and taking into account the effect of various forces acting on it. Some of these forces are gravitational, molecular dispersion, double-layer interaction, and drag forces and torques. If this trajectory takes the particle to the pore surface, then the condition for its deposition is inspected. If the attractive forces are larger than the repulsive ones, then the particle deposits on the surface. But, if the pore surface is perfectly smooth, the particle will simply roll on it and leaves the pore. In reality the pore surface is not smooth, therefore it is necessary to include a measure of surface roughness in the model. Imdakm and Sahimi did this by distributing overhangs of various heights on the pore surface. If the particle does deposit, its location is recorded. The fluid flow exerts a drag force on the deposited particle, which increases the resistance to the flow in that pore. This increase in the resistance is equivalent to a decrease in the effective radius of the pore. After each deposition, the pore radius is updated. Moreover, particle deposition affects the flow field in the network. Hence, each time a few particles are deposited on the pore surfaces, the flow field in the entire network is recalculated, from which the permeability of the pore space at that time is obtained.

If the particle is not deposited, it travels through the pore, arrives at a node, and selects its next pore with a probability proportional to the flow rate in that pore. The motion of the particles is biased by the flow field in

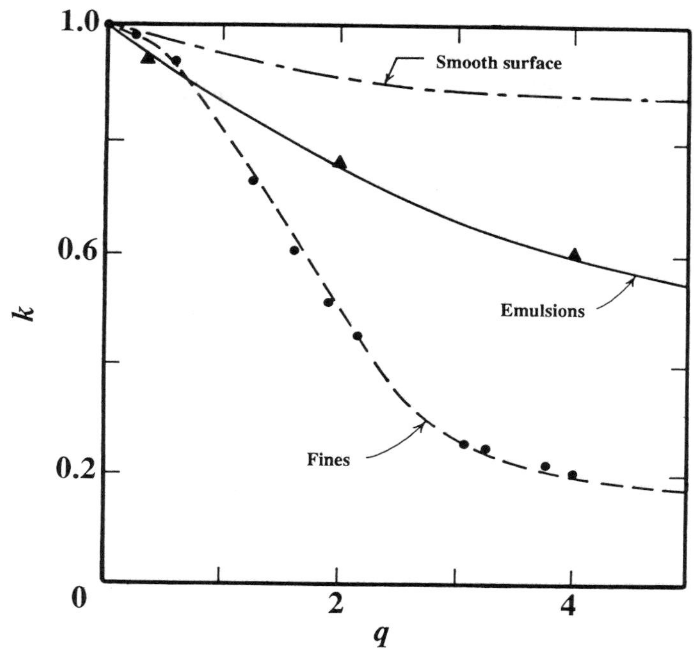

Figure 8.7 *Comparison of percolation predictions of the variations of the permeability k of porous media with the injected pore volume q during flow of fines and emulsion droplets with the experimental data (symbols). k has been normalized by its value at $q = 0$ (after Sahimi and Imdakm 1991).*

favor of the pores that carry a larger flux of fluid. This is similar to hydrodynamic dispersion in Chapter 6. If the effective radius of the particles is *larger* than the radius of the pore, the pore is completely plugged by the particle. Otherwise, the trajectory of the particle within the new pore is calculated and the process continues. If deposition of the emulsions is modeled, only monolayer formation is allowed. Figure 8.7 compares experimental data with the predicted permeabilities for fines migration and for flow of stable emulsions; the agreement is excellent. Also shown are the predicted permeabilities for a smooth pore surface. The permeability of the porous medium obeys a scaling law near the percolation threshold that is different from (2.11) for random percolation. Complete details can be found in the original papers of the authors.

8.7 Conclusions

Reaction, deposition, and transport processes in a porous medium that cause dynamical changes in the morphology of the medium are percolation processes. Percolation allows us to develop microscopic models for a large class of such phenomena, and to make quantitative predictions for their various properties. Given the success of such models, we may use them for designing better structures for our porous media, such as catalyst particles, so they can better resist the dynamical changes in their structures and have a much longer useful life time.

References

Arbabi, S. and Sahimi, M., 1991, *Chem. Eng. Sci.* **46**, 1739.
Beeckman, J.W., Nam, I.S. and Froment, G.F., 1987, *Stud. Surf. Sci. Catal.* **34**, 365.
Beyne, A.O.E. and Froment, G.F., 1990, *Chem. Eng. Sci.* **45**, 2089.
Brenner, H. and Gajdos, L.J., 1977, *J. Colloid Interface Sci.* **58**, 312.
Cai, M., Edwards, B.F., and Han, H., 1991, *Phys. Rev. A* **43**, 656.
Englman, R., Jaeger, Z., and Levi, A., 1984, *Phil. Mag.* **B50**, 307.
Gavalas, G.R., 1981, *Combust. Sci. Tech.* **24**, 197.
Gyune, M.F. and Edwards, B.F., 1992, *Phys. Rev. Lett.* **68**, 2692.
Hughes, R., 1984, *Catalyst Deactivation*, New York, Academic Press.
Imdakm, A.O. and Sahimi, M., 1991, *Chem. Eng. Sci.* **46**, 1977.
Jackson, R., 1977, *Transport in Porous Catalysts*, Amsterdam, Elsevier.
Kerstein, A.R. and Bug, A.L.R., 1986, *Phys. Rev. B* **34**, 1754.
Kerstein, A.R. and Edwards, B.F., 1987, *Chem. Eng. Sci.* **42**, 1629.
Kerstein, A.R. and Niksa, S., 1984, in *Twentieth Symposium (International) on Combustion*, Pittsburgh, the Combustion Institute, p. 941.
Melkote, R.R. and Jensen, K.F., 1989, *Chem. Eng. Sci.* **44**, 649.
Mohanty, K.K., Ottino, J.M., and Davis, H.T., 1982, *Chem. Eng. Sci.* **37**, 905.
Nam, I.S. and Froment, G.F., 1987, *J. Catal.* **108**, 271.
Petersen, E.E., 1957, *AIChE J.* **3**, 443.
Reyes, S. and Jensen, K.F., 1986, *Chem. Eng. Sci.* **41**, 333.
Reyes, S. and Jensen, K.F., 1987, *Chem. Eng. Sci.* **42**, 565.
Sadakata, M., Yoshino, A., Sakai, T., and Matsuoka, H., 1984, in *Twentieth Symposium (International) on Combustion*, Pittsburgh, the Combustion Institute, p. 913.
Sahimi, M., 1991, *Phys. Rev. A* **43**, 5367.
Sahimi, M., Gavalas, G.R., and Tsotsis, T.T., 1990, *Chem. Eng. Sci.* **45**, 1443.
Sahimi, M. and Imdakm, A.O., 1991, *Phys. Rev. Lett.* **66**, 1169.
Sahimi, M. and Tsotsis, T.T., 1985, *J. Catal.* **96**, 552.
Sahimi, M. and Tsotsis, T.T., 1987, *Phys. Rev. Lett.* **59**, 888.
Sahimi, M. and Tsotsis, T.T., 1988, *Chem. Eng. Sci.* **43**, 113.
Shah, N. and Ottino, J.M., 1987a, *Chem. Eng. Sci.* **42**, 63.
Shah, N. and Ottino, J.M., 1987b, *Chem. Eng. Sci.* **42**, 73.
Sotirchos, S.V. and Zarkanitis, S., 1993, *Chem. Eng. Sci.* **48**, 1487.
Sundback, C.A., Beer, J.M., and Sarofim, A.F., 1984, in *Twentieth Symposium (International) on Combustion*, Pittsburgh, the Combustion Institute, p. 1495.
Tien, C. and Payatakes, A.C., 1979, *AIChE J.* **25**, 737.
Yortsos, Y.C. and Sharma, M.M., 1986, *AIChE J.* **32**, 46.
Yu, H.-C. and Sotirchos, S.V., 1987, *AIChE J.* **33**, 382.

9
Fractal diffusion and reaction kinetics

9.0 Introduction

Chapter 8 discusses the application of percolation to transport and reaction in porous media. In this chapter we consider the effect of a percolating structure on diffusion and reaction kinetics. Diffusion in a disordered system is strongly influenced by its structure. Our main goal in this chapter is to study the effect of a *fractal* structure on diffusion, an extreme but rather common form of disorder. Reaction kinetics in systems as diverse as catalytic materials, membranes, and molecular aggregates can also be influenced by their structure and, if the reactions are diffusion-controlled, the effect of disorder is amplified. The reactions that we consider here are mostly *heterogeneous*, i.e., they take place at the interface between two different phases, e.g., a gas and a solid phase, as in porous catalysts. The key difference between this chapter and the previous chapter is that, the diffusion–reaction processes in Chapter 8 cause dynamical changes in the structure of the media in which they take place, whereas the processes in this chapter do not cause such changes. Instead, they are greatly affected by the structure of the media, and their behavior shows dramatic departure from the classical laws of diffusion and reaction. Moreover, there are many systems in which a *nonchemical* reaction takes place. For example, excitation recombination and quenching in photosynthetic units are in this class of nonchemical reactions. Depite their fundamental difference with chemical reactions, many features of the two phenomena are the same.

9.1 Fractal diffusion in percolation systems

Chapters 5 and 6 show that diffusion can be simulated by a simple random walk. A random walker is put on an occupied site of the network, which at each time step takes one step to one of the nearest neighbor occupied sites of the node at which it is currently residing. If the bond conductances are distributed, then the probability of taking a step from site i to site j is proportional to the conductance of the bond ij. If all open bonds have the same conductance, then the walker selects a nearest neighbor occupied site with probability $1/Z$, where Z is the coordination number. This basic idea was first used by Brandt (1975) to study diffusion of noble gasses in glasses, modeled as a percolation network. However, the idea was popularized by de Gennes (1976), who made an analogy between the motion of the random walker in the percolation network and that of an ant in a labyrinth, followed by the numerical simulations of Mitescu and Roussenq (1976). For the most recent works see Roman (1990) and Sahimi and Stauffer (1991). For recent reviews see Havlin and Ben-Avraham (1987) and Haus and Kehr (1987).

One of the most important properties of the random walk is the average probability $\langle P(r,t)\rangle$ that the diffusing particle is at point $\mathbf{r}$ at time t, where $r = |\mathbf{r}|$, and the averaging is over all initial positions of the diffusing particles. For ordinary diffusion in a d-dimensional and macroscopically homogeneous but microscopically disordered system, $\langle P(r,t)\rangle$ is given by the Gaussian distribution

$$\langle P(r,t)\rangle \sim t^{-d/2} \exp\left(-\frac{r^2}{4D_e}t\right), \tag{9.1}$$

and obeys the classical diffusion equation. The mean-square displacement $\langle r^2\rangle$ is given by $\langle r^2\rangle = 2dD_e t$, where the effective diffusivity D_e is constant and independent of time. Since $\langle P(r,t)\rangle$ is proportional to the concentration of the diffusing particles (see Chapter 6), these are just another way of expressing Fick's law of diffusion. We recall the precise conditions under which (9.1) is valid. If $r_s = \langle r^2\rangle^{1/2}$, then (9.1) is valid if r_s is much larger than the length scale over which the system is macroscopically homogeneous. This length scale for percolation systems is the correlation length ξ_p, and therefore only for $r_s >> \xi_p$ one has Gaussian diffusion with a constant diffusivity. From (9.1) it also follows that the probability of being at the origin at time t, $\langle P_0\rangle = \langle P(0,t)\rangle$, is given by

$$\langle P_0\rangle \sim t^{-d/2}. \tag{9.2}$$

However, as the percolation threshold p_c is neared ξ_p becomes large, and therefore at least at the initial stages of diffusion r_s will be smaller than ξ_p. At p_c the correlation length is divergent, the largest percolation cluster is a fractal object, hence diffusion cannot be Gaussian at *any* time. When $r_s \ll \xi_p$, $\langle r^2\rangle$ grows *nonlinearly* and *subdiffusively* with the time t such that

$$\langle r^2\rangle \sim t^{2/d_w}, \tag{9.3}$$

where d_w is called the fractal dimension of the random walk (Chapter 6). A transport process characterized by (9.3) is called *anomalous* (Gefen *et al.* 1983) or *fractal* diffusion (Sahimi *et al.* 1983). Gaussian diffusion is just a special case of fractal diffusion when $d_w = 2$. Equation (9.3) is valid for *all* fractal systems, for which one always has, $d_w > 2$. The implication of (9.3) is that, $D_e = d\langle r^2\rangle/dt$ is time-dependent and *vanishes* as $t \to \infty$. Hydrodynamic dispersion studied in Chapter 6, and turbulent diffusion, are two examples where d_w can be *smaller* than two.

Let us now define a new quantity d_s by

$$d_s = 2\frac{D_f}{d_w}, \tag{9.4}$$

where D_f is the fractal dimension of the system; for the largest percolation cluster $D_f = D_c$, as given by (2.16). This quantity is called the *spectral* or *fracton* dimension, a term that was first used by Alexander and Orbach (1982). One motivation for introducing this quantity was to obtain $\langle P_0\rangle$ for a fractal medium. Diffusion in a fractal medium is said to be *recurrent*, i.e., given enough time the random walker will visit most points of the medium. Thus, $\langle P_0\rangle$ should be proportional to $1/V$, where V is the volume that the random walker explores. Since, $V \sim (\langle r^2\rangle)^{D_f/2}$, we obtain, using (9.3), $\langle P_0\rangle \sim t^{-d_s/2}$. Another motivation for introducing this quantity is discussed in Chapter 10, but let us note that in some sense d_s is a measure of the interplay between the fractal geometry of the system (measured by D_f) and the dynamics of diffusion (measured by d_w). For nonfractal (homogeneous) media, $d_w = 2$, $D_f = d$, and $d_s = d$. This tells us immediately that the role of d_s for fractal systems is similar to that of d for Euclidean systems, and we shall see that this is indeed the case.

For percolation networks we can relate d_w to the other exponents defined in Chapter 2. Suppose that diffusion takes place only on the sample-spanning cluster. Then, from (2.10) we have, $D_e \sim (p - p_c)^{\mu - \beta_p} \sim \xi_p^{-\theta}$, where $\theta = (\mu - \beta_p)/\nu_p$. Since we are interested in the fractal diffusion regime where $r_s \ll \xi_p$, we can replace ξ_p with $\langle r^2\rangle^{1/2}$ and write, $D_e \sim \langle r^2\rangle^{-\theta/2}$. On the other hand, $D_e \sim d\langle r^2\rangle/dt \sim \langle r^2\rangle^{-\theta/2}$, which after integration yields $\langle r^2\rangle \sim t^{2/(2+\theta)}$, implying that (Gefen *et al.* 1983)

$$d_w = 2 + \frac{\mu - \beta_p}{\nu_p}, \tag{9.5}$$

which shows that $d_w > 2$ (since $\mu > \beta_p$). Equation (9.5) also tells us that, in order to estimate μ, we can calculate d_w by a simple random walk (Pandey *et al.* 1984). This idea turned out to be a rather accurate method of estimating μ, although not as accurate as once hoped. More accurate methods are now available. Random walk simulations can be vectorized for use in supercomputers such as the Cray Y-MP, which allows us to study diffusion in very large systems (see Sahimi and Stauffer 1991, where the computer program for doing this is described and given). Equation (9.5) can be rewritten as, $d_w = 2 - d + \mu/\nu_p + D_c$, which means that $d_w > D_c$, and therefore $d_s < 2$. If we use the values of the critical exponents (Chapter 2) we find that, $d_s(d = 2) \simeq d_s(d = 3) \simeq 1.32$, and $d_s(d \geqslant 6) = 4/3$. Alexander and Orbach (1982) conjectured that $d_s = 4/3$ at *all* dimensions. If this conjecture were true, it would have provided a simple relation between μ, which is a dynamic exponent, and β_p and ν_p, which are static (geometrical) exponents. It is now believed this conjecture is wrong, but $d_s \simeq 4/3$ is still an excellent approximation for percolation clusters and many other fractal systems.

When does a crossover from fractal to Gaussian diffusion take place? Because for $r_s \ll \xi_p$ diffusion is fractal, while it is Gaussian for $r_s \gg \xi_p$, the crossover time t_{co} is when $r_s \sim \xi_p$. Since diffusion time is of the order of r_s^2/D_e, we must have, $t_{co} \sim \xi_p^2/D_e \sim \xi_p^{d_w}$. Using the scaling laws for ξ_p and D_e, (2.6) and (2.10), we obtain

$$t_{co} \sim (p - p_c)^{-2\nu_p - \mu + \beta_p}. \tag{9.6}$$

But $-2\nu_p - \mu + \beta_p$ is always negative, which means that as p_c is approached the crossover time t_{co} becomes very large, and we have to wait a very long time to observe Gaussian diffusion. This has an important implication for experimental studies of diffusion in percolating systems, e.g., a porous medium near its percolation threshold. If D_e is measured at $t < t_{co}$, it will be dependent upon the measurement time, whereas for $t >> t_{co}$ it will be a constant. At p_c the crossover time t_{co} is divergent, which explains why diffusion is always fractal at p_c.

For fractal diffusion the probability $\langle P(r, t)\rangle$ is no longer given by (9.1), and does not obey the classical diffusion equation. The proper form of $\langle P(r, t)\rangle$ for diffusion on fractals was a controversial subject for several years. It now appears that the following equation

$$\langle P(r, t)\rangle \sim t^{-d_s/2} \exp\left[-\left(\frac{r}{\langle r^2(t)\rangle^{1/2}}\right)^u\right], \tag{9.7}$$

originally suggested by Guyer (1985), can provide accurate representation of diffusion on fractals, where

$$u = \frac{d_w}{d_w - 1}. \tag{9.8}$$

Equation (9.7) is *not* in general exact, but it is a very good approximation; see Havlin and Ben-Avraham (1987) for a discussion of this. Equation (9.7) also tells us that for fractal systems

$$\langle P_0\rangle \sim t^{-d_s/2}, \tag{9.9}$$

derived above using a scaling argument. A comparison of (9.9) with (9.2) confirms our assertion that, for fractal media, d_s plays the same role as d for Euclidean systems.

What happens if the random walk is performed over *all* percolation clusters (as in, e.g., a diffusion experiment in a porous medium in which the the molecules diffuse in both the isolated and connected regions of the pore space)? Chapter 2 shows that at p_c there is a distribution of clusters with a wide variety of sizes. We need to average $\langle r^2 \rangle$ over the cluster size distribution (2.14). This was done by Gefen *et al.* (1983), who showed that diffusion is still fractal but with $\langle r^2 \rangle \sim t^{2/d_w^a}$, where a denotes an averaging over all clusters, and $d_w^a = 2d_w/(2 - \beta_p/\nu)$. Moreover, since only the sample-spanning cluster is fractal, but the collection of all clusters is not fractal, (9.9) is no longer valid. We need to average (9.9) over all clusters whose distribution is given by (2.16). Because we have finite clusters as well as the sample-spanning (infinite) cluster, $\langle P_0 \rangle$ will approach a constant for long enough times, and we obtain

$$\langle P^a(0,t) - P^a(0,\infty) \rangle \sim t^{-d_s^a/2}, \tag{9.10}$$

where, $d_s^a = 2d/d_s$. Nonfractality of all clusters causes D_c to be replaced by d.

9.2 Diffusion-controlled reactions on percolation clusters

Before discussing various diffusion-controlled reactions, we introduce a key quantity, $S(t)$, which is *the number of distinct sites visited* by a random walker. What is the physical meaning of $S(t)$? Since $S(t)$ counts the number of distinct points that have been reached by the random walker, $dS(t)/dt$, the number of distinct sites visited per unit of time, or the volume of the system explored per unit of time (if we give a unit volume to each site), is in some sense a measure of the *efficiency* of the random walker for reaching various regions of the system (de Gennes 1983). For example, if reacting molecules are diffusing thoughout a system, their ability to reach one another and react, i.e., the reaction rate, should be related to $dS(t)/dt$. For diffusion in homogeneous media, $S(t)$ and $\langle P_0(t) \rangle$ are related through their Laplace transforms (Montroll and Weiss 1965)

$$\hat{S}(\lambda) = \frac{1}{\lambda^2} \langle \hat{P}_0(\lambda) \rangle, \tag{9.11}$$

where λ is the Laplace transform variable conjugate to t. Using (9.2) we immediately obtain

$$S(t) \sim \begin{cases} t^{1/2} & d = 1 \\ t \ln t & d = 2 \\ t & d \geq 3 \end{cases} \tag{9.12}$$

Self-similarity of fractal media means that (9.11) cannot be expected to hold at short or intermediate times, because the effect of the origin of the random walk is strong at such times. But for sufficiently long times, we expect (9.11) to be applicable to fractal media, and using (9.9) we obtain for $d_s < 2$

$$S(t) \sim t^{d_s/2}, \tag{9.13}$$

which was first given by Rammal and Toulouse (1983). We now consider several types of diffusion-controlled reactions in disordered media, and discuss the main results and their experimental realization.

9.2.1 Diffusion-controlled trapping

Consider the reaction

$$A + B \rightarrow B \tag{9.14}$$

where A is a diffusing reactant with concentration C_A, and B is a stationary absorbing reactant or trap with concentration C_B, distributed randomly throughout the system. We would like to know how C_A varies with the time t. This problem was of great interest for a long time, as it is directly relevant to excition trapping and recombination in disordered materials (for a review see Blumen *et al.* 1986). It has been shown that the long-time behavior of C_A is dominated by the diffusion of A into the regions that contain no B at all, which are very rare. Using this, the following relation has been derived (see, e.g., Donsker and Varadham 1975, Grassberger and Procaccia 1982, Kayser and Hubbard 1983)

$$C_A \sim \exp[-aC_B^{2/(d+2)} t^{d/(d+2)}], \tag{9.15}$$

where a is a constant. If the B molecules are distributed in a fractal medium, then (9.15) would still be valid provided that we replace d with d_s. Since for percolation $d_s \simeq 4/3$, we obtain

$$C_A \sim \exp(-aC_B^{3/5} t^{2/5}). \tag{9.16}$$

However, since C_A is controlled by rare events, experimental confirmation of (9.16) is very difficult if not impossible.

9.2.2 Diffusion-controlled annihilation

Consider next the reaction

$$A + A \to 0, \tag{9.17}$$

where molecules A, initially distributed randomly in the system, diffuse until they collide with each other and disappear. In classical kinetics this reaction is considered as second order, which means that the rate of reaction R is given by

$$R = KC_A^2, \tag{9.18}$$

where K is a constant. On the other hand, the rate of reaction is, by definition, the rate of change of C_A with respect to time, i.e., $R = -dC_A/dt$, which, when substituted into (9.18) and integrated, yields

$$\frac{1}{C_A} - \frac{1}{C_{A0}} = Kt, \tag{9.19}$$

where C_{A0} is the initial concentration of A. Equation (9.19) tells us that the dimensionality of the system plays no role in this reaction. This is similar to a mean field treatment and, indeed, implicit in the derivation of (9.19) is that the system can be represented by a *mean* concentration. This is true if the system is *well mixed*, i.e., if C_A is uniform everywhere. However, if this uniformity cannot be achieved, then (9.19) is not expected to be valid, at least for low-dimensional systems, because the fluctuations of C_A throughout the system cannot be ignored. For this case, Kang and Redner (1985) showed that

$$C_A = C_{A0}t^{-d/2}, \tag{9.20}$$

which is valid for $d \leqslant 2$. For $d \geqslant 2$, we have $C_A \sim t^{-1}$, which is essentially the same as (9.19).

What happens if (9.17) is carried out in a fractal system? From our discussions so far it is easy to guess that

$$C_A \sim t^{-d_s/2} \sim \frac{1}{S(t)}, \tag{9.21}$$

which is valid for $d_s < 2$. Since for percolation, $d_s \simeq 4/3$, (9.21) takes on a simple form, $C_A \simeq t^{-2/3}$, and simulations confirmed this (Meakin and Stanley 1984). Equation (9.21) also implies that the rate coefficient K is no longer a constant. In this case we write (de Gennes 1983)

$$K \sim \frac{dS(t)}{dt}, \tag{9.22}$$

which, together with (9.13) yields

$$K \sim t^{d_s/2-1}. \tag{9.23}$$

Note that K is a constant only if $d_s = 2$. If we now use $d_s \simeq 4/3$ for percolation systems, we obtain

$$K \sim t^{-1/3} \sim C_A^{1/2}. \tag{9.24}$$

Equations (9.21) and (9.24) have been confirmed experimentally. Here we discuss these experiments briefly and refer the reader to Kopelman (1988) for their full details. In one experiment an exciton fusion reaction was studied in which two triplet excitations fused and produced a singlet excitation. The reaction takes place inside a mixed crystal of naphthalene alloy, made of $C_{10}H_8$ and $C_{10}D_8$. The naphthalene molecules are distributed randomly among the alloy's lattice sites, and the excitons are restricted to the $C_{10}H_8$ clusters. If the mole fraction x of $C_{10}H_8$ is not large enough, the excitons are restricted to small clusters and cannot diffuse very far. But if x is larger than the percolation threshold x_c, in this case $x_c \simeq 0.08$, then the excitons can explore a large cluster. At or very close to x_c we expect (9.24) to hold, whereas for $x \gg x_c$, we expect to obtain the classical result that K is independent of time. Figure 9.1 shows the results of the experiment for $x_c \simeq 0.08$, i.e., at the percolation threshold of the naphthalene clusters, and the slope of the curve is 0.32 ± 0.03, in complete agreement with (9.24). Another interesting experiment which confirmed (9.24) is naphthalene photodimerization in porous membranes. This is, a reaction of the type,

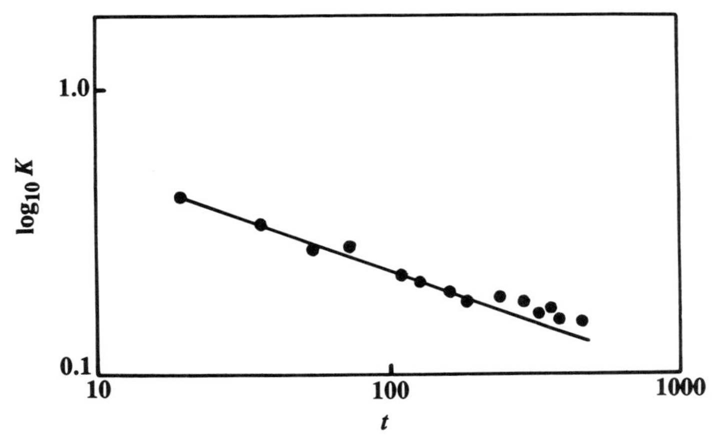

Figure 9.1 *Logarithmic plot of the reaction rate K versus time t (in ms) for the exciton fusion reaction in isotopically mixed naphthalene crystals at the percolation threshold. The slope of the line is* $d_s/2 - 1 = -0.32 \pm 0.03$ *(after Kopelman 1988).*

$A^* + A^* \rightarrow A^{**} \rightarrow A + A + a$ *photon*, where A^* is a naphthalene molecule excited to its first triplet state, and A^{**} is the intermediate dimer in its first excited singlet state. The experiments were carried out in a solution embedded in many types of porous membranes. The slope of the $\ln K$ versus $\ln t$ plot was again found to be about 1/3, in good agreement with (9.24).

What happens if reaction (9.17) is carried out under *steady state* conditions, where C_A does not depend on time? In this case classical kinetics does not have much to offer: it still tells us that the reaction is second order. But suppose that reaction (9.17) is carried out at steady state in a fractal system. If we write

$$R = -\frac{dC_A}{dt} = KC_A^2 = K_0 C_A^2 t^{d_s/2 - 1}, \tag{9.25}$$

where K_0 is a constant, and integrate this equation we obtain

$$\frac{1}{C_A} - \frac{1}{C_{A0}} = \frac{2K_0}{d_s} t^{d_s/2}. \tag{9.26}$$

Thus, as $C_A \rightarrow 0$ (or as $t \rightarrow \infty$), we have

$$t \sim \left(\frac{d_s}{2K_0}\right)^{2/d_s} C_A^{2/d_s}, \tag{9.27}$$

which, when substituted into (9.25), yields

$$R = K_0 C_A^n, \tag{9.28}$$

$$n = 1 + \frac{2}{d_s}, \tag{9.29}$$

so that with $d_s \simeq 4/3$, we obtain $n = 5/2$, distinctly different from the classical value, $n = 2$. Equation (9.29) implies that if the reaction is carried out in a one-dimensional system ($d_s = 1$), e.g., a molecular thin wire, its order would be *three*, although the reaction is still bimolecular. Moreover, there are fractal systems whose fractal dimension is *less* than one. Such systems were called *dust fractals* by Mandelbrot (1982). A practical example may be catalytic islands on noncatalytic support, which are in fact quite common. For dust fractals, we have $0 < d_s < 1$, and hence, $3 < d_s < \infty$, i.e., we may have a reaction with a very large order. Equation (9.29) has also been confirmed experimentally (see Kopelman 1988).

Consider now the diffusion-controlled reaction

$$A + B \rightarrow 0, \tag{9.30}$$

where A and B are initially distributed randomly throughout the system. Classical kinetics tells that

$$R = -\frac{dC_A}{dt} = -\frac{dC_B}{dt} = KC_AC_B. \tag{9.31}$$

If $c = C_{B0} - C_{A0}$, then, the solution of (9.31) is easily found to be

$$C_A(t) = \frac{c}{(1 + c/C_{A0})\exp(Kct) - 1}, \tag{9.32}$$

whereas if $C_{A0} = C_{B0}$, then,

$$C_A(t) = \frac{C_{A0}}{1 + KC_{A0}t}. \tag{9.33}$$

Similar to reaction (9.17), while concentration fluctuations are important to reaction (9.30), especially at low dimensions, (9.32) and (9.33) do not take them into account. That is, even if reaction (9.30) takes place in a nonfractal, but low-dimensional system, we may still observe drastic deviations from the predictions of classical kinetics. For this case, Kang and Redner (1985) presented the following scaling argument to obtain the relation between C_A and t. Assume that initially $C_{A0} = C_{B0} = C_0$, and consider a region of linear size L. In this region there are $C_0L^d \pm [C_0L^d]^{1/2}$ A molecules (and similarly for B), where the second term represents the fluctuations in the concentration due to the initial random distribution of the molecules. After a time $t \sim L^2/D_m$, where D_m is the microscopic diffusivity, all pairs of the molecules will be consumed, and we will be left with only about $(C_0L^d)^{1/2}$ molecules of A or B, so that the concentration of A or B will be, $C_A \sim (C_0L^d)^{1/2}/L^d$, which means that

$$C_A = C_0t^{-d/4}, \tag{9.34}$$

which is valid for $d \leqslant 4$. Only when $d = 4$, do the predictions of (9.33) and (9.34) become comparable. The same type of argument can be used for deriving (9.20). To obtain the corresponding expression for a fractal system, all we have to do is replace L^d with L^{D_f} and $t \sim L^2$ with $t \sim L^{d_w}$. This results in

$$C_A = C_0t^{-d_s/4}, \tag{9.35}$$

which means that d is simply replaced with d_s. Equation (9.35) also implies that for percolation systems

$$C_A = C_0t^{-1/3}. \tag{9.36}$$

These results were also confirmed by numerical simulations (Kang and Redner 1985, Meakin and Stanley 1984).

9.3 Conclusions

The main conclusion of this chapter is that concentration fluctuations can play a dominant role in diffusion-controlled reactions. Percolation-like disorder amplifies such fluctuations, and therefore gives rise to unusual kinetics that cannot be predicted by the classical and mean field theories.

References

Alexander, S. and Orbach, R., 1982, *J. Physique Lett.* **43**, L625.

Blumen, A., Klafter, J., and Zumofen, G., 1986, in *Optical Spectroscopy of Glasses*, edited by I. Zschokke, Dordrecht, Reidel, p. 199.

Brandt, W.W., 1975, *J. Chem. Phys.* **63**, 5162.

de Gennes, P.G., 1976, *La Recherche* **7**, 919.

de Gennes, P.G., 1983, *C. R. Acad. Sci. Paris* **296 II**, 881.

Donsker, M.D. and Varadham, S.R.S., 1975, *Commun. Pure Appl. Math.* **28**, 525.

Gefen, Y., Aharony, A., and Alexander, S., 1983, *Phys. Rev. Lett.* **50**, 77.

Grassberger, P. and Procaccia, I., 1982, *J. Chem. Phys.* **77**, 6281.

Guyer, R.A., 1985, *Phys. Rev. A* **32**, 2324.

Haus, J.W. and Kehr, K.W., 1987, *Phys. Rep.* **150**, 263.

Havlin, S. and Ben-Avraham, D., 1987, *Adv. Phys.* **36**, 695.

Kang, K. and Redner, S., 1985, *Phys. Rev.* **A32**, 435.

Kayser, R.F. and Hubbard, J.B., 1983, *Phys. Rev. Lett.* **51**, 79.

Kopelman, R., 1988, *Science* **241**, 1620.

Mandelbrot, B.B., 1982, *The Fractal Geometry of Nature*, San Francisco, Freeman.

Meakin, P. and Stanley, H.E., 1984, *J. Phys. A* **17**, L173.

Mitescu, C. and Roussenq, J., 1976, *C. R. Acad. Sci. Paris* **283A**, 999.

Montroll, E.W. and Weiss, G.H., 1965, *J. Math. Phys.*

Pandey, R.B., Stauffer, D., Margolina, A., and Zabolitzky, J.G., 1984, *J. Stat. Phys.* **34**, 877.

Rammal, R. and Toulouse, G., 1983, *J. Physique Lett.* **44**, 13.

Roman, H.E., 1990, *J. Stat. Phys.* **58**, 375.

Sahimi, M., Hughes, B.D., Scriven, L.E., and Davis, H.T., 1983, *J. Chem. Phys.* **78**, 6849.

Sahimi, M. and Stauffer, D., 1991, *Chem. Eng. Sci.* **46**, 2225.

10
Vibrations and density of states of disordered materials

10.0 Introduction

Disordered and rigid materials can have unusual properties. Consider, for example, gel polymers that are of great practical importance. The gelling solution at the gel point, i.e., the point at which there is a transition from a liquid-like system (sol) to a solid-like system (gel), has unique properties: Its viscosity is *infinite*, whereas all of its elastic moduli are *zero*. Moreover, the gel network has a fractal structure at the gel point. Chapter 11 shows that the gel point is in fact a percolation threshold. Another example is carbon-black composites, which are made out of rigid particles in a soft matrix. The rigid components form very large fractal clusters that resemble those near the percolation threshold p_c, much like the gel networks. If we try to deform such materials, the distribution of the forces that are exerted on the monomers of the gel network, or on the rigid and soft particles of the carbon-black composites, depends on the topology of the system, and if the system is fractal, this distribution will not follow classical laws of mechanics, just as flow, diffusion, and reaction kinetics in fractal media do not follow the usual classical laws (Chapter 9). The same applies to the vibrational properties of such materials. In the previous chapters, the effect of fractal and percolation structures on diffusion, flow, and reaction kinetics are discussed. In the present chapter we focus on vibrations of rigid percolation networks, discuss the effect of a fractal structure on the vibrational properties, and review their experimental realization.

10.1 Vibrations and density of states of homogeneous rigid structures

Consider an $L \times L \times L$ network in which each site contains a particle of mass m. The particles are connected to each other by springs. Suppose that $\mathbf{u}_i = (u_{ix}, u_{iy}, u_{iz})$ is the displacement of the particle at i. Then, the equation of the motion for this particle at time t is given by Newton's law

$$m \frac{\partial^2 \mathbf{u}_i}{\partial t^2} = \sum \mathbf{F}, \tag{10.1}$$

where $\mathbf{F}$ is any kind of force that is acting on the particle. If the springs that connect the particles are harmonic and can only tolerate stretching forces, then $\mathbf{F}$ is given by Hooke's law (force = spring constant × displacement), and the equation of motion becomes

$$m \frac{\partial^2 \mathbf{u}_i}{\partial t^2} = \sum_{\langle ij \rangle} k_{ij}[(\mathbf{u}_j - \mathbf{u}_i) \cdot \mathbf{R}_{ij}], \tag{10.2}$$

where $\mathbf{R}_{ij}$ is a unit vector from i to j, k_{ij} is the spring (elastic) constant of the spring between i and j, and the sum is over all particles j that are connected to i. For simplicity we take $m = 1$. Of course we can include other types of forces in (10.2), e.g., the bond-bending or angle-changing forces that are discussed in Chapters 4 and 11.

The standard method of analyzing (10.1) for vibrational properties is to assume, $\mathbf{u}_i = \mathbf{A}_i \exp(-i\omega t)$, where ω is the frequency of vibrations, $\mathbf{A}_i$ is an unknown vector to be determined, and $i = \sqrt{-1}$. Substituting this into (10.1) yields a set of $n = L^3$ simultaneous linear equations for the $\mathbf{A}_i$ which has n positive eigenvalues $\omega_1^2, \omega_2^2, \ldots,$ and n eigenvectors $\mathbf{A}_{e1}, \mathbf{A}_{e2}, \ldots$. Then, the solution to $\mathbf{u}_i$ is given by

$$\mathbf{u}_i = Re\left[\sum_j c_j \mathbf{A}_{ej} \exp(-i\omega_j t)\right], \tag{10.3}$$

where the c_j are complex numbers that have to be determined from the initial conditions, and Re denotes the real part of the complex number. Having determined the $\mathbf{u}_i$, we can obtain all vibrational properties of the system. One of the most important properties is $N(\omega)$, called the *vibrational density of states*. $N(\omega)d\omega$ is the number of vibrational modes with a frequency between ω and $\omega + d\omega$. This is an important quantity for obtaining the specific heat and thermal conductivity of the system, and can itself be obtained from the distribution of the eigenmodes ω_i. But there is a much simpler way of obtaining the dependence of $N(\omega)$ on ω.

Consider, for example, (10.2) and take its Fourier transform

$$-\omega^2\tilde{\mathbf{u}}_i = \sum_{\langle ij\rangle} k_{ij}[(\tilde{\mathbf{u}}_j - \tilde{\mathbf{u}}_i)\cdot \mathbf{R}_{ij}], \tag{10.4}$$

where $\tilde{\mathbf{u}}_i(\omega)$ is the Fourier transform of $\mathbf{u}_i$. If we write (10.4) for one principal direction of the network, say x, we obtain

$$-\omega^2\tilde{u}_{ix} = \sum_{\langle ij\rangle} k_{ij}(\tilde{u}_{jx} - \tilde{u}_{ix}). \tag{10.5}$$

Consider now the diffusion equation in discretized form (5.9)

$$\frac{\partial P_i}{\partial t} = \sum_{\langle ij\rangle} W_{ij}[P_j(t) - P_i(t)], \tag{10.6}$$

where $P_i(t)$ is the probability of finding a diffusing particle at site i of the network at time t, and W_{ij} is the transition rate, i.e., the probability that the diffusing particle jumps from site i to site j per unit time (or the conductance of the bond between i and j). If we take the Laplace transform of (10.6), and ignore the term that arises from the initial condition, we obtain

$$\lambda\hat{P}_i(\lambda) = \sum_{\langle ij\rangle} W_{ij}(\hat{P}_j - \hat{P}_i), \tag{10.7}$$

where $\hat{P}_i(\lambda)$ is the Laplace transform of $P_i(t)$, and λ is the Laplace transform variable conjugate to t. If we now compare (10.5) and (10.7), we see that they are very similar: The role of $\tilde{u}_{ix}$ is played by $\hat{P}_i$, and that of $-\omega^2$ by λ. We may then interpret λ as the frequency for diffusion, just as ω is the frequency for vibrations.

Consider now $\langle P_0(t)\rangle$, the average probability of being at the origin at time t (see Chapter 9), where the averaging is taken over all initial positions. Equation (9.2) tells us that for *macroscopically homogeneous* media we have

$$\langle P_0(t)\rangle \sim t^{-d/2}, \tag{10.8}$$

where d is the dimensionality of the system. $N(\omega)$ should in principle be obtained from $\tilde{\mathbf{u}}(\omega)$. But it can be shown that $N(\omega)$ and $\langle P_0(t)\rangle$ are related to each other through the following equation (Alexander *et al.* 1978)

$$N(\omega) = -\frac{2\omega}{\pi}\, Im\,\langle \hat{P}_0(-\omega^2)\rangle, \tag{10.9}$$

where *Im* denotes the imaginary part of the complex number. If we take the Laplace transform of (10.8), and substitute it in (10.9), we obtain

$$N(\omega) \sim \omega^{d-1}, \tag{10.10}$$

a well-known result. Equation (10.10) is valid at low frequencies (i.e., long wavelengths or large length scales over which the system is homogeneous), such that $\omega < \omega_{co}$, where ω_{co} is some cutoff or crossover frequency. It can be shown that even if we obtain $N(\omega)$ from the solution of (10.1), it would still follow (10.10), provided that the system is macroscopically homogeneous. This is also clear in (10.10). There is nothing in this equation that tells us whether $N(\omega)$ was obtained from the solution of (10.1) or (10.6). Only the dimensionality of the system has entered this equation. Vibrational states in homogeneous systems, expressed by (10.10), are usually called *phonons*.

10.2 Vibrations and density of states of fractal and percolation networks

What happens if our network is fractal? For example, suppose that we have a percolation network with $L \ll \xi_p$, where ξ_p is the correlation length defined in Chapter 2. Then, the sample-spanning cluster is a fractal object with a fractal dimensionality D_c. Our network is no longer macroscopically homogeneous, so we do not expect (10.10) to hold. The density of states of fractal networks was first discussed by Alexander and Orbach (1982). If $L \ll \xi_p$, then (9.9) tells us that

$$\langle P_0(t)\rangle \sim t^{-d_s/2}, \tag{10.11}$$

where d_s is the spectral dimension introduced and discussed in Chapter 9. At first we may be tempted to use (10.11) in (10.9) to obtain the corresponding expression for $N(\omega)$ for fractal networks. But we cannot do this straightforwardly. To see this, recall that $d_s = 2D_c/d_w$ then substitute $D_c = d - \beta_p/\nu_p$ (2.16) and $d_w = 2 + (\mu - \beta_p)/\nu_p$ (9.5) in this equation, to obtain

$$d_s = 2\frac{\nu_p d - \beta_p}{2\nu_p + \mu - \beta_p}. \tag{10.12}$$

Observe that the exponents ν_p and β_p are purely topological properties and do not depend on the equation governing a given phenomenon in a percolation network, but μ is the exponent for conductivity and diffusivity of percolation networks, both governed by (9.6). On the other hand, recall that in a percolation network of springs near p_c, the effective elastic moduli $G(p)$ of the system obey the following scaling law (2.9)

$$G(p) \sim (p - p_c)^f, \tag{10.13}$$

where in general $f \neq \mu$. Therefore, in this case the nature of the equation from the solution of which $N(\omega)$ is to be extracted, or more precisely, the nature of the forces that are exerted on the particles of our network, is important. If we wish to take into account the true vector nature of the vibrational density of states (important if the system is fractal) then we are forced to introduce a new quantity d_{es}, called *the elastic spectral dimension* and defined by replacing μ by f in (10.12). According to Webman and Grest (1985)

$$d_{es} = 2\frac{\nu_p d - \beta_p}{2\nu_p + f - \beta_p}. \tag{10.14}$$

We are also forced to define *two* different densities of states for the vibrations of the sample-spanning cluster in the fractal regime. $N(\omega)$ is the density of states when the governing equation is scalar, such as (10.6); using (10.11) in (10.9) this yields

$$N(\omega) \sim \omega^{d_s - 1}. \tag{10.15}$$

The second quantity is the *elastic density of states* $N_e(\omega)$ which, in analogy with (10.15), is given by

$$N_e(\omega) \sim \omega^{d_{es} - 1}. \tag{10.16}$$

Equations (10.15) and (10.16) are valid for high frequencies (i.e., short wavelengths or short length scales), such that $\omega > \omega_{co}$. Vibrational states on fractal structures, expressed by (10.15), were called *fractons* (phonons on fractals) by Alexander and Orbach (1982).

What is the cutoff frequency at which a crossover between (10.10) on one hand and (10.15) (10.16) on the other takes place? Consider, for example, the crossover between (10.10) and (10.15). In this case, the crossover between phonons and fractons is similar to the crossover between normal and fractal diffusion discussed in Chapter 9; it takes place at a time scale t_{co} such that (9.6)

$$t_{co} \sim (p - p_c)^{-2\nu_p - \mu + \beta_p} \sim (p - p_c)^{-\nu_p d_w}. \tag{10.17}$$

This time scale corresponds, in the Laplace transform space, to a crossover value $1/\lambda_{co}$. Since $\omega_{co}^2 \sim \lambda_{co}$, we obtain

$$\omega_{co} \sim (p - p_c)^{\nu_p + (\mu - \beta_p)/2} \sim (p - p_c)^{\nu_p d_w/2}, \tag{10.18}$$

which is the equation derived by Alexander and Orbach (1982) using a different analysis. We can rewrite (10.18) in terms of the percolation correlation length ξ_p, the result is

$$\omega_{co} \sim \xi_p^{-d_w/2} \sim \xi_p^{-D_c/d_s}. \tag{10.19}$$

Chapter 2 discusses finite-size scaling; for an $L \times L \times L$ system such that $L \ll \xi_p$, (10.19) is equivalent to

$$\omega_{co} \sim L^{-d_w/2} \sim L^{-D_c/d_s}, \tag{10.20}$$

so that as L increases (but $L/\xi_p \ll 1$), the cutoff or crossover frequency decreases. Alternatively, we can convert (10.20) into an equation for a cutoff length scale L_{co} at which a crossover between (10.10) and (10.15) takes place

$$L_{co} \sim \omega_{co}^{-d_s/D_c}. \tag{10.21}$$

With the help of (10.18) we can write (10.10) and (10.15) in a unified form

$$N(\omega) \sim \omega^{d_s-1} h(\omega/\omega_{co}), \tag{10.22}$$

where $h(x)$ is a scaling function. Since for $\omega << \omega_{co}$ we want to recover (10.10) we must have $h(x) \sim x^{d-d_s}$ for $x << 1$. On the other hand, we also want to recover (10.15) in the limit $\omega >> \omega_{co}$, which means that $h(x) \sim$ constant for $x >> 1$. This also implies that for $L >> \xi_p$ we have

$$N(\omega \sim (p-p_c)^{\nu_p d_w (d_s-d)/2} \omega^{d-1} \sim \omega_{co}^{d_s-d} \omega^{d-1}. \tag{10.23}$$

Equation (10.23) implies that the ratio of (10.15) and (10.23) is a *constant* at $\omega = \omega_{co}$. The quantity $N(\omega)$ can also be calculated by an effective-medium approximation discussed in Chapter 5 (Sahimi 1984, Derrida *et al.* 1984), since the Green function G_0 defined by (5.11) is just the Laplace transform of $\langle P_0(t)\rangle$. The EMA results are, $N(\omega) \sim \omega_{co}^{-d/2}\omega^{d-1}$ for the phonon regime, and $N(\omega) =$ *constant* for the fraction regime. Equations (10.15) and (10.16) are valid for the sample-spanning cluster. If we are interested in the vibrational density of states for *all* clusters, then (10.15) and (10.16) can still be used, except that D_c should be replaced with d, because only the sample-spanning cluster is a fractal object for $L << \xi_p$, and the collection of all clusters is space filling and not fractal.

The difference between (10.15) and (10.16) is dramatic. Chapter 9 shows that for *both* two- and three-dimensional percolation networks at p_c, $d_s \simeq 4/3$ which implies that

$$N(\omega) \sim \omega^{1/3}. \tag{10.24}$$

If $N(\omega)$ decreases with ω, then the medium is mechanically stable. To understand this, recall that small frequencies imply larger length scales; if over such length scales $N(\omega)$ decreases, the implication would be that there are fewer and fewer vibrational modes, i.e., the system is mechanically stable. On the other hand, if we use the value of f for three-dimensional percolation systems in which both stretching and bond-bending forces are important (see Chapter 11), $f \simeq 3.75$, we obtain $d_{es} \simeq 0.87$ and (10.16) implies that

$$N_e(\omega) \sim \omega^{-0.13}, \tag{10.25}$$

As ω decreases there are larger and larger number of vibrating modes. This implies that such systems *may not become too large*, because if they do, they might lose their mechanical stability, have to restructure themselves, and cross over to another structure. If only central forces are important (10.2) then (see Chapter 11) $f \simeq 2.1$, and $N(\omega) \sim \omega^{0.28}$, almost identical with (10.24). But we should keep in mind that while (10.24) is valid for fractons in a scalar-like system, a network of springs with central forces is still a vector system.

Is there a crossover between (10.15) and (10.16)? Feng (1985) argued that in a percolation network in which stretching and bond-bending forces are both present, another length scale L_{bb}, in addition to ξ_p, is important. If $L_{bb} >> \xi_p$, then at low frequencies (10.10) and at higher frequencies (10.15) governs the density of states, with a crossover at ω_{co}. If $L_{bb} << \xi_p$ low frequencies dictate the phonon regime until a characteristic frequency ω_{bb} is reached at which there is a crossover from (10.10) to (10.16). When the higher frequency ω_{co} is reached, there is a second crossover from (10.16) to (10.15). From (10.19) we guess that

$$\omega_{bb} \sim \xi_p^{-D_c/d_{es}}. \tag{10.26}$$

According to Feng, the characteristic length scale L_{bb} is given by

$$L_{bb} \sim (\beta/\alpha)^{1/2}, \tag{10.27}$$

where α and β are respectively, the stretching and bond-bending force constants, discussed in Chapters 4 and 11.

10.3 Experimental verification

Over the past several years many authors have investigated the validity and generality of the above results. For example, Yakubo *et al.* (1990) carried out extensive simulations and calculated the density of states of fractal structures to check the validity of (10.15) in two and three dimensions. They found that to a very good degree of accuracy (10.15) holds. Rammal *et al.* (1984) carried out random walk simulations on the sample-spaning cluster of percolation networks for $2 \leqslant d \leqslant 6$, and found that $d_s \simeq 4/3$, an indirect confirmation of (10.15). As usual, we are interested in the experimental verification of such results. Here we discuss a few experimental studies that support the above theoretical results. The reader should consult Courtens *et al.* (1989) and Kjems (1991) for more complete discussions of past work. Most of these studies involve aerogels, so we discuss briefly their preparations and properties.

Aerogels are highly porous solid materials that have a very tenuous structure. Their porosity can be as high as 99%, so they often have unique or very unusual properties. For example, they can be made in transparent form, they have very small thermal conductivity, and because of their large porosity they possess large internal surface area. Their properties give them a wide range of applications, from catalyst supports to thermal insulators and radiators. They can be prepared by a variety of methods using different materials, but silica aerogels have received the widest attention. They are produced by hydrolysis of $Si(OR)_4$, where R represents either CH_3 or C_2H_5. A catalyst, either an acid or a base, is also used; this strongly influences the reaction. The degree of hydrolysis is controlled by the ratio $[Si(OR)_4]/[H_2O]$, and the final density of the aerogel is controlled by $[Si(OR)_4]/[ROH]$, where [. . .] denotes concentration. Because of the acidic or basic catalyst, the pH of the solution also has a strong effect on the structure of the gel. Hydrolysis produces -SiOH groups which then polymerize into -Si-O-Si-. Particles start to grow in the liquid solution and after some time form a gel network. The solvent is removed to obtain the solid porous structure. Aerogels are obtained if the solvent is removed at a temperature *above* its critical point. A more complete discussion of aerogel preparation is given by Courtens *et al.* (1989).

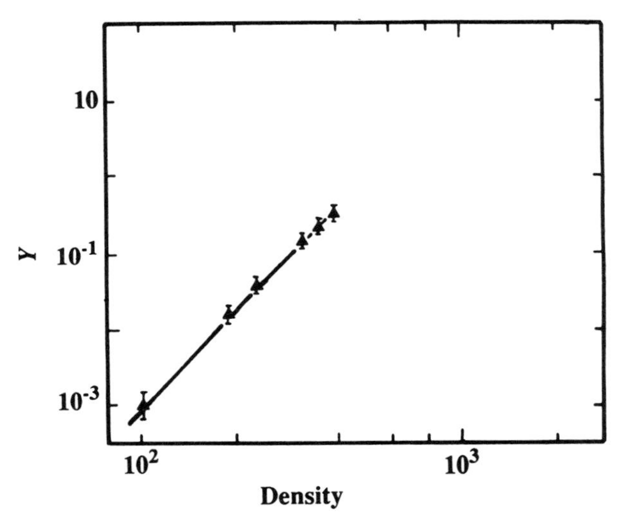

Figure 10.1 *Young modulus Y of silica aerogels gels (in GPa) versus their apparent density (in* kg m^{-3}*) (after Woignier et al. 1988).*

Silica aerogels appear to have the structure of percolation networks. Evidence for this comes from two different directions. Woignier *et al.* (1988) measured the Young modulus Y of silica aerogels as a function of density and volume fraction. Figure 10.1 shows their results. Over nearly one order of magnitude, Y scales with the volume fraction ε of the solid as

$$Y \sim (\varepsilon - \varepsilon_c)^f, \tag{10.28}$$

where ε_c is the critical volume fraction or the percolation threshold of the network. As this figure indicates, $f \simeq 3.8$, in excellent agreement with the three-dimensional elastic percolation networks discussed in Chapters 2, 4, and 11. The other evidence comes from the measurement of the fractal dimension of the gel. One way of determining the fractal dimension is through small-angle neutron scattering (SANS). If we define a correlation function $C(r)$ by

$$C(r) \sim \sum_{r'} s(r')s(r + r'), \tag{10.29}$$

where $s(r) = 1$, if a point at a distance r from the origin belongs to the network, and $s(r) = 0$ otherwise, then the scattering intensity $I(q_s)$ is the Fourier transform of $C(r)$, with $q_s = 4\pi \sin(\theta_s/2)/\xi$, where θ_s is the scattering angle and ξ is the wavelength. For fractal structures with a fractal dimensionality D_f, $C(r) \sim r^{D_f - d}$ and therefore for $q_s\xi >> 1$ we have

$$I(q) \sim q_s^{-D_f}. \tag{10.30}$$

Figure 10.2, taken from Vacher *et al.* (1988), presents typical data obtained by SANS. The fractal dimension of silica aerogels was found to be, $D_f \simeq 2.45$, very close to the fractul dimension of three-dimensional sample-spanning percolation clusters, $D_c \simeq 2.5$ (see Chapter 2). It seems that at least one class of silica aerogels are essentially percolation fractals.

The next step is to verify the scaling law for the density of states, (10.15) or (10.16). One way of doing this is by plotting the crossover or cutoff frequency ω_{co} as a function of q_s. Since q_s is inversely proportional to the length scale L, (10.20) tells us that

$$\omega_{co} \sim q_s^{D_f/d_s}. \tag{10.31}$$

This plot should yield, $\omega_{co} \sim q_s^{1.88}$ if silica aerogels are percolation fractals and $d_s \simeq 4/3$. Figure 10.3, taken from Courtens and Vacher (1989), shows such a plot for a series of aerogels with various densities. The data were obtained by Brillouin scattering of visible light and indicate that, $\omega_{co} \sim q_s^{1.9}$, in very good agreement with the theoretical prediction. The density of states

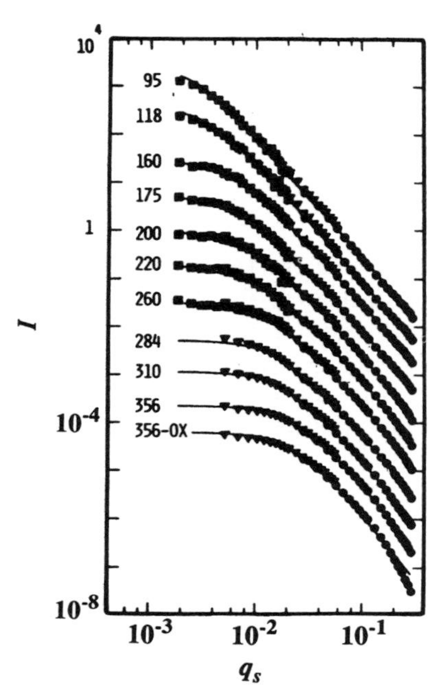

Figure 10.2 *Scattering intensities (relative to H_2O) for 11 silica aerogel samples. The top 10 samples are unreacted and neutral, whereas the bottom one is an oxidized sample. The density of the samples increases from top to bottom. q_s is in $Å^{-1}$ (after Vacher et al. 1988).*

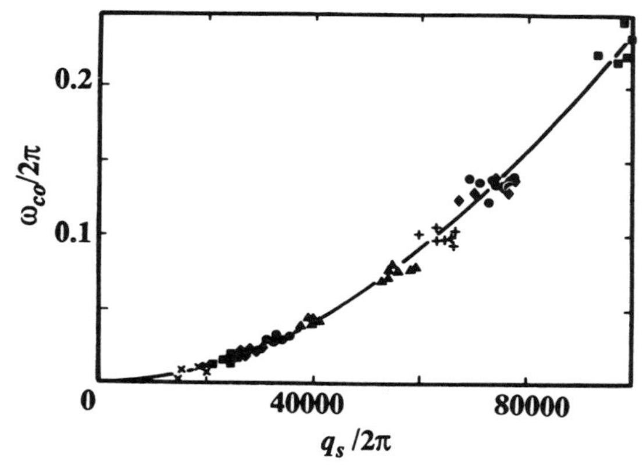

Figure 10.3 *The fracton dispersion curve obtained from the Brillouin determination of the cutoff frequency ω_{co} and scattering vector q_s. Symbols are the same as in Fig. 10.2.*

itself can also be measured directly by incoherent neutron scattering. In this method, protons are chemically bonded to the surface of the particles in the network. Because protons have a large cross-sectional area, they can dominate incoherent scattering from a network that contains a small number of hydrogen atoms, such as the aerogels. On the other hand, deuterons have a small cross section. From the difference between the scattering from two different samples in which small amounts of OD and OH have been bonded to the sample, the density of states can be measured. This was done by Vacher *et al.* (1989), who obtained $N(\omega) \sim \omega^{0.85 \pm 0.15}$, which does not agree with (10.15) or (10.16). This difference may be explained in various ways. One is that the networks used in this study are not simply sample-spanning percolation clusters. In addition to the main gel network, many other smaller clusters are also grown. When the solvent is removed, we have a *distribution* of clusters. We are not measuring the incoherent scattering from just one percolation cluster, but from a collection of them. Thus, we should have $N(\omega) \sim \omega^{2d/d_w - 1}$, rather than the usual $N(\omega) \sim \omega^{2D_f/d_w - 1}$, because the collection of all percolation clusters is not fractal, only the sample-spanning cluster is fractal. This reasoning predicts $N(\omega) \sim \omega^{0.6}$, reasonably close to the experimental measurements. Vacher *et al.* gave a rather different reason for this difference. They argued that the finite clusters attach themselves to the sample-spanning cluster, making it denser, thereby increasing the effective value of d_s from its theoretical value of about 4/3 to a larger value of 1.85. A recent computer simulation study by Nakanishi (1993) using perculation networks supports this. We may also argue that energy resolution in the experiment was not sufficient to yield enough accuracy. Indeed, the experiments of Schaefer *et al.* (1990), using more accurate techniques, yielded $d_s \simeq 1.22 \pm 0.14$, again consistent with the theoretical expectation.

Finally, the crossover between (10.15) and (10.16) was studied in the experiments of Vacher *et al.* (1990) using silica aerogels. They measured the density of states of the gels and found that at low frequencies their data can be fitted with $d_{es} \simeq 0.9$, in good agreement with the theoretical expectation discussed above. At higher frequencies the data can be fitted with $d_s \simeq 1.7 \pm 0.2$, once again higher than $d_s \simeq 4/3$. It is not yet clear whether this discrepancy can be explained by the same type of reasoning discussed above. For the most recent discussion of this see Alexander *et al.* (1993).

10.4 Conclusions

Fractal properties play a central role in the vibrational properties of disordered materials. Vibrational properties do not follow the classical laws derived for homogeneous media. The result, $d_s \simeq 4/3$, obtained from percolation networks, seems to explain many sets of experimental data for various materials. It points to the crucial role of percolation in such materials.

References

Alexander, S., Bernasconi, J., and Orbach, R., 1978, *J. Physique Colloq.* **39(C6)**, 706.
Alexander, S. and Orbach, R., 1982, *J. Physique Lett.* **43**, L625.
Alexander, S., Courtens, E., and Vacher, R., 1993, *Physica A* **195**, 286.
Courtens, E. and Vacher, R. 1989, *Proc. Roy. Soc. London* **A423**, 55.
Courtens, E., Vacher, R., and Stoll, E., 1989, *Physica D* **38**, 41.
Derrida, B., Orbach, R., and Yu, K.W., 1984, *Phys. Rev. B* **29**, 6645.
Feng, S., 1985, *Phys. Rev. B* **32**, 5793.
Kjems, J.K., 1991, in *Fractals and Disordered Systems*, edited by A. Bunde and S. Havlin, Berlin, Springer.
Nakanishi, H., 1993, *Physica A* **196**, 33.
Rammal, R., Angles d'Auriac, J. C., and Benoit, A., 1984, *Phys. Rev. B* **30**, 4087.
Sahimi, M., 1984, *J. Phys. C* **17**, 3957.
Schaefer, D.W., Brinker, C.J., Richter, D., Farago, B., and Frick, B., 1990, *Phys. Rev. Lett.* **64**, 2316.
Vacher, R., Woignier, T., Pelous, J., and Courtens, E., 1988, *Phys. Rev. B* **37**, 6500.
Vacher, R., Woignier, T., Pelous, J., Coddens, G., and Courtens, E., 1989, *Europhys. Lett.* **8**, 161.
Vacher, R., Courtens, E., Coddens, G., Heidemann, A., Tsujimi, Y., Pelous, J., and Foret, M., 1990, *Phys. Rev. Lett.* **65**, 1008.
Webman, I. and Grest, G., 1985, *Phys. Rev. B* **31**, 1689.
Woignier, T., Phalippou, J., and Pelous, J., 1988, *J. Phys. France* **49**, 289.
Yakubo, K., Courtens, E., and Nakayama, T., 1990, *Phys. Rev. B* **42**, 8933.

11
Structural, mechanical, and rheological properties of branched polymers and gels

11.0 Introduction

Two key works in the early 1970s demonstrated clearly how polymers can be studied both theoretically and experimentally. de Gennes (1971) showed that there is a close connection between linear polymers (i.e., those whose monomers have functionality $Z = 2$) and a statistical mechanical model, namely, the n-vector model. If no two polymer parts occupy the same point in space, the linear polymer corresponds to the limit $n \to 0$. If this restriction can be ignored, the polymer corresponds to a random walk or the $n = -2$ limit of the model. This discovery enabled us to apply modern methods of statistical mechanics, such as renormalization group theory, to the study of linear polymers. On the experimental side, Cotton (1974) used small-angle neutron scattering, and a labeling technique in which the hydrogen atoms along the polymer chain were replaced by deuterium, to show that it is possible to detect one polymer chain among many others in a solution. Thus, we can analyze a single polymer chain and compare the experimental results with the theoretical predictions.

The works of de Gennes and Cotton were restricted to linear polymers, i.e., those in which each monomer is connected to two neighboring monomers. However, if we use monomers with functionality $Z > 2$, where each monomer is connected to up to Z neighboring monomers, then at least two other classes of polymers can be obtained. If the reaction time t is relatively short and below, but close to, a characteristic time t_g, then we obtain branched polymers in the solution, usually called a *sol*, that form a viscous solution. These branched polymers are large but finite clusters of monomers. On the other hand, if the reaction time is larger than t_g, a very large solid network of connected monomers appears that is usually called a *chemical gel*, or simply a gel. The gel network has interesting structural, mechanical, and rheological properties. The characteristic time t_g is called the *gelation time*, and the point at which the gel network appears for the first time is called the *gel point* (GP). Familiar such sol–gel transformations may be the milk-to-cheese transition, pudding, gelatine, etc. This chapter models various properties of the sol and gel phases, especially near GP. Gelation is the phase transition from the sol phase to the gel phase; it has been described by a percolation model.

11.1 Percolation model of polymerization and gelation

Consider a solution of molecules or monomers with functionality $Z \geqslant 3$. To understand the connection with percolation, suppose that the monomers occupy the sites of a periodic lattice. With probability p, two nearest-neighbor monomers (sites) can react and form a chemical bond between them. If p is small, only small polymers are formed. As p increases, larger and larger polymers (clusters of connected monomers) with a broad size distribution are formed. This mixture of clusters of reacted monomers and the isolated unreacted monomers represents the sol phase. For $p > p_c$, where p_c is a characteristic value that depends on Z (or, the number of nearest-neighbors of a monomer of the lattice), an "infinite" cluster of reacted monomers is formed. This cluster represents the gel network discussed above. Near GP the gel usually coexists with a sol such that the finite polymers are trapped in the interior of the gel. As $p \to 1$, almost all monomers react, and the sol phase disappears completely. Thus, p_c signals a *connectivity* transition: for $p > p_c$, an infinite cluster exists and the system is mainly a solid gel (with possibly a few finite polymers). The fraction of chemical bonds formed at GP (related to the fraction of reacted monomers at GP) is the analog of the bond percolation threshold p_{cb} defined in Chapter 2. Therefore, the formation of branched polymers and gels is very similar to a percolation process. In reality, the monomers do not react with each other randomly, and there are usually some correlations between the reaction of monomers with one another, but this does not change the main results of this chapter.

This model of polymerization and gelation was essentially invented by Flory (1941) and Stockmayer (1943), who were interested in the formation of large branched polymers. However, Flory and Stockmayer considered what we call percolation on a Bethe lattice, a branching, but loopless, structure (see Chapter 2). Although they did not use the terminology of percolation processes, it is now recognized that the Flory–Stockmayer theory represents the *mean field* limit of percolation. Stauffer (1976) and de Gennes (1976) were first to recognize the relevance of percolation on three-dimensional lattices to the critical behavior of branched polymers and gels.

Physical gels, as opposed to chemical gels, are formed when no *permanent* chemical reaction takes place between the particles or monomers: Only a *reversible association* links the particles to each other. In physical gels, monomers are usually nonreactive small solid particles. They are formed when the solid particles are somehow attached to each other and form long chains. Two examples: silica aerogels and structures formed during the gelation of silica particles in pure water or NaCl solutions. Now we describe some important properties of sol and gel phases and their percolation modeling.

11.2 Mechanical, rheological, and structural properties of branched polymers and gels

What is the signature of the sol–gel phase transition? Experimental studies of sol–gel transitions usually proceed by measuring the time evolution of the rheological or mechanical properties (e.g., viscosity or elastic moduli) during the chemical reaction leading to gelation, assuming that the experimental parameter, time or frequency, and the theoretical one, the number of cross-links, are linearly related in the vicinity of GP. Rheological measurements are usually performed by using a cone and plate rheometer or by the more accurate magnetic sphere rheometer. The ranges of shear rates, deformations, and times of measurements of these devices allow the determination of the steady state zero-shear viscosity η and the steady state linear elastic moduli G up to the *vicinity* of the phase transition at GP, but it has proven to be almost impossible to do such measurements *at* GP.

All the experimental data for the elastic moduli of gel networks above GP, but close to GP indicate that the elastic moduli G of the network obey a scaling law given by

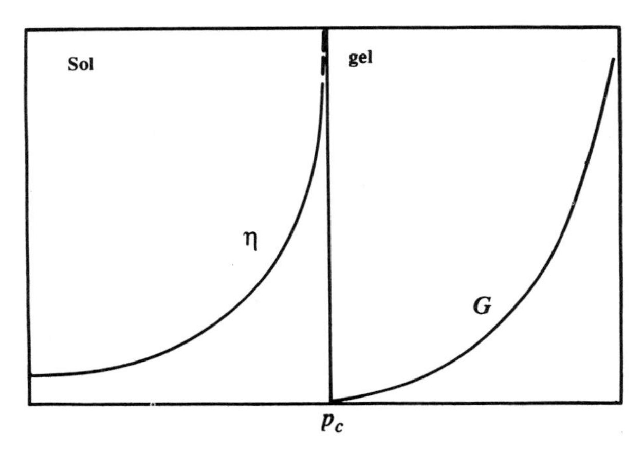

Figure 11.1 *Typical variations of viscosity η and elastic moduli G with the fraction of reacted monomers p during gelation.*

$$G \sim (p - p_c)^f, \quad p > p_c. \tag{11.1}$$

On the other hand, the viscosity of the sol phase diverges as

$$\eta \sim (p_c - p)^{-k}, \quad p < p_c. \tag{11.2}$$

Typical variations of η and G with p are shown in Fig. 11.1. In practice, it is precisely this divergence of η that signals the formation of a gel network. Experimental determination of k is more difficult than determination of f. The major obstacles are that accurate determination of GP is often difficult and, moreover, the measurements cannot be made *at* GP and in the limit of *zero frequency*. To estimate k we usually define a frequency-dependent complex modulus $G^*(\omega) = G'(\omega + iG''(\omega)$ at frequency ω, where $i = \sqrt{-1}$. At GP and for low frequencies, we have

$$G' \sim G'' \sim \omega^\Delta, \tag{11.3}$$

with

$$\Delta = \frac{f}{f+k}, \tag{11.4}$$

where G' (storage modulus) and G'' (loss modulus) describe storage and dissipation in an oscillating strain field of constant amplitude. Typical variations of G' and G'' with ω are shown in Fig. 11.2 for a polycondensed gel very close to GP.

The complex moduli $G^*(\omega)$ is sometimes written as $G^* = G + i\omega\eta$, for which Durand *et al.* (1987) proposed that

$$G^*(\omega, \varepsilon) \sim \varepsilon^f h_1(i\omega\varepsilon^{f+k}), \tag{11.5}$$

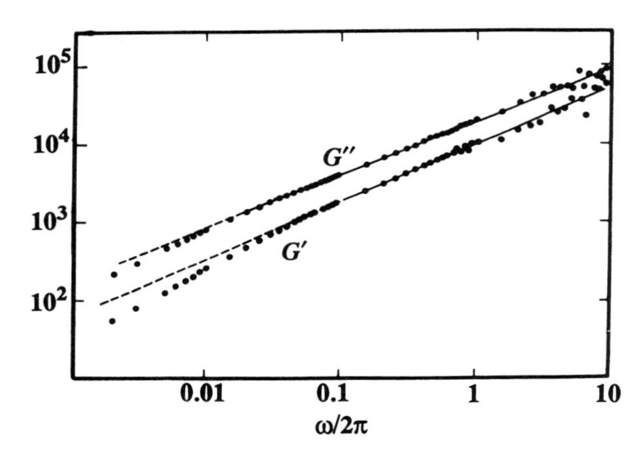

Figure 11.2 *Frequency dependence of the storage modulus G′ and loss modulus G″ for a polycondensed gel close to the gel point (after Durand et al. 1987).*

where $\varepsilon = |p - p_c|$, and $h_1(x)$ is a scaling function. The significance of this scaling equation is that it allows us to collapse the data for *all* values of ε and ω onto a single curve, usually called the *master curve* by polymer researchers. In the low-frequency regime, we do not expect G^* to depend on ε, and depend only on ω. Then we find that $G^* \sim \omega^{-i\Delta}$, equivalent to (11.3). Moreover, there is a loss angle δ defined by $\tan\delta = G'/G''$. The remarkable property of δ is that at GP it takes on a value δ_c given by

$$\delta_c = \frac{\pi}{2}(1 - \Delta) = \frac{\pi}{2}\frac{k}{f + k}. \tag{11.6}$$

so that, if the critical exponents f and k are universal, the loss angle δ_c will also be universal.

An important problem in polymerization and gelation is the determination of GP, either for avoiding it to prevent gelation so that a polymer with certain properties can be processed, or for making materials very close to GP, since such materials have unusual properties. GP depends on the functionality of the polymer. Chapter 2 shows that percolation thresholds decrease with increasing coordination numbers, which are the analog of polymer functionality. Thus, polymers with cross-links of high functionality gel very early. Holly *et al.* (1988) proposed using the loss angle δ for locating GP. They argued that since as GP is reached $\tan(\delta)$ becomes independent of the frequency (11.6), the intersection of the various curves in a plot of $\tan(\delta)$ versus time at various frequencies should give the location of GP. Figure 11.3, taken from Lin *et al.* (1991), shows how this method is used for locating the GP for a physical gel.

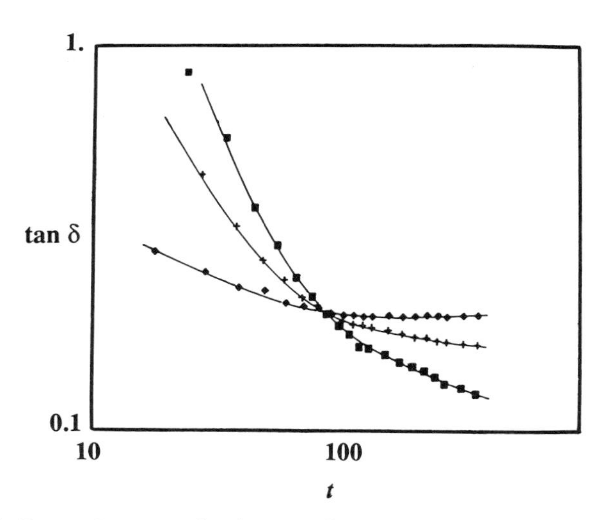

Figure 11.3 *Determination of gel point from data for loss angle δ. Time t is in minutes. The data correspond to frequencies 31.6 rad s⁻¹ (diamonds), 1.0 rad s⁻¹ (+), and 0.0316 rad s⁻¹ (squares) (after Lin et al. 1991).*

Viscosity and elastic moduli are rheological and mechanical properties of branched polymers and gels. They characterize the dynamics of the polymerization. We may measure, directly or indirectly, the distribution of relaxation times $H(t)$ in the reaction bath. The moments are directly related to the viscosity and the elastic moduli. To obtain $H(t)$, we define a complex viscosity $\eta^*(\omega) = G^*(\omega)/(i\omega)$, which is related to the distribution of relaxation times by (see, e.g., Ferry 1980)

$$\eta^*(\omega) = \int \frac{H(t)}{1 + i\omega t} dt. \tag{11.7}$$

Then, using (11.5), we can back-calculate $H(t)$ (Daoud 1988)

$$H(t) \sim t^{-\Delta} h_2(t\varepsilon^{f+k}), \tag{11.8}$$

where h_2 is another scaling function. This equation indicates that in the scaling regime near GP, $H(t)$ is a slowly decaying power law. Daoud (1988) calculated *two* distinct average or characteristic times:

$$T_1 = \int H(t)dt \Big/ \int [H(t)/t]dt \sim \varepsilon^{-k} \sim \eta, \tag{11.9}$$

and

$$T_2 = \int tH(t)dt \Big/ \int H(t)dt \sim \varepsilon^{-f-k} \sim \frac{\eta}{G}. \tag{11.10}$$

Note that T_2 is the *longest* characteristic time of the system. The existence of the distribution of relaxation times and its scaling form given by (11.8) means that *any* relaxation property in the intermediate time or frequency range is *not* exponential, but follows a *power law*. The divergence of T_1 and T_2 is responsible for the fact that measurements of η or G fail at GP, because steady state conditions cannot be reached in a finite time.

There are also several important structural (static) properties of branched polymers and gel networks that can be measured directly or indirectly. The *gel fraction* $GF(p)$ is the fraction of the monomers that belong to the gel network. It is obvious that $GF(p) > 0$ only if $p > p_c$, and that GF is the analog of percolation fraction or percolation probability $P(p)$ defined in Chapter 2. Of particular interest to us is the behavior of $GF(p)$ near p_c. It is found in this region that

$$GF(p) \sim (p - p_c)^{\beta}, \tag{11.11}$$

completely similar to (2.3). The gel fraction can be measured by simply weighing the solid gel at different times during polymerization. The correlation or connectivity length ξ of branched polymers diverges as p_c is approached, according to the scaling law

$$\xi \sim |p - p_c|^{-\nu}, \tag{11.12}$$

which is the analog of (2.6). Above GP the correlation length of polymers can be interpreted as the mesh size of the gel network. For any length scale greater than ξ the gel network is essentially homogeneous. Below GP the correlation length is the typical radius of the polymers in the sol phase. In this case, those polymers whose radii are much larger than ξ are described by a different set of critical exponents. *At* GP the gel network is *not* homogeneous but is a self-similar fractal object with a fractal dimension D_p that in d-dimensions is given by

$$D_p = d - \frac{\beta}{\nu}, \tag{11.13}$$

the same as (2.16). The number distribution of polymers, i.e., the probability $Q(n, \varepsilon)$ that a polymer of the sol phase contains n monomers at a distance ε from GP, can also be defined. This quantity is the analog of n_s, defined in Chapter 2. Thus, in analogy with (2.14) we can write

$$Q(n, \varepsilon) \sim n^{-\tau} h_3(\varepsilon n^{\sigma}), \tag{11.14}$$

where h_3 is another scaling function. Using this distribution, we can define two distinct mass averages. The *weight average molecular weight* is given by

$$M_w = \frac{\int n^2 Q(n, \varepsilon)\, dn}{\int n Q(n, \varepsilon)\, dn} \sim \varepsilon^{-\gamma}. \tag{11.15}$$

The quantity M_w is the analog of $S_p(p)$, the mean size of percolation clusters defined by (2.2), where the sum in that equation has been replaced here by an integral. In the polymer literature M_w is also called the *degree of polymerization*. The second mass average is defined by

$$M_z = \frac{\int n^3 Q(n, \varepsilon) dn}{\int n^2 Q(n, \varepsilon) dn} \sim \varepsilon^{-1/\sigma}, \tag{11.16}$$

where $\sigma = (\tau - 2)/\beta$, as in percolation (see Chapter 2). The existence of two distinct mass averages is similar to the existence of two distinct average relaxation times defined above. But the average $\langle M \rangle = \int n Q(n, \varepsilon) / \int Q(n, \varepsilon)$ does *not* diverge at GP. We are now in a position to compare measurements of various polymer properties with the predictions of percolation theory. Since one of the main predictions of percolation is the existence of universal critical exponents and fractal dimensions, and since the numerical values of any polymer property are not universal and depend on the microscopic structure of the polymer, we focus here on a comparison between the measured universal exponents and fractal dimensions and the predictions of percolation.

11.3 Comparison of experimental data for the structural properties of branched polymers and gels with the percolation predictions

After a polymer is formed by a chemical reaction, the experimentalist usually analyzes its structure by diluting it in a good solvent. Such branched polymers in a dilute solution of a good solvent may swell and have a radius *larger* than their extent at the end of the chemical reaction. Thus, it is important to consider typical polymers and swollen polymers. Let us now discuss the experimental evidence for the applicability of a percolation model for describing the structural properties of gels and the structural properties of branched polymers in a dilute solution of a good solvent. Consider first a swollen branched polymer in a good solvent whose radius is *larger* than the polymer correlation length ξ. Its structural properties are described by *lattice animals*, which are percolation clusters *below* the percolation threshold whose radii are *larger* than the percolation correlation length ξ_p. Although lattice animals have a close connection with percolation, their scaling properties are *different* from those of percolation clusters. To see this, let us first define a few key properties of lattice animals. Let $A_s(p)$ be the average number (per lattice site) of clusters and a_{sm} be the total number of geometrically different configurations for a cluster of s sites and perimeter m. Thus, $A_s(p) = \sum_m a_{sm} p^s (1-p)^m$. For large s, the asymptotic behavior of $A_s(p)$ is described by the power law

$$A_s(p) \sim \lambda_g^s s^{-\theta}, \qquad (11.17)$$

where θ is a universal exponent independent of the coordination number of the lattice, whereas the *growth parameter* λ_g is not universal. Moreover, if n_a is the number of monomers (sites) in a lattice animal of radius R_a, then for large values of n_a, a fractal dimension D_a is defined by

$$n_a \sim R_s^{D_a}. \qquad (11.18)$$

Lubensky and Isaacson (1978) and Family and Coniglio (1980) showed that the exponents θ and D_a are *not* related to any of the percolation exponents defined in Chapter 2. Moreover, Parisi and Sourlas (1981) showed that

$$\theta = \frac{d-2}{D_a} + 1, \qquad (11.19)$$

and that

$$D_a = 2, \quad d = 3. \tag{11.20}$$

There are two other differences that distinguish lattice animals from percolation clusters. The exponents θ and D_a defined above are valid for *any* $p < p_c$ (remember that the percolation exponents are defined for $p \simeq p_c$). All we require is $R_a >> \xi_p$. The upper critical dimension for lattice animals is eight, two more than for percolation. The upper critical dimension is the dimension at which the mean field approximation to the critical exponents becomes exact.

We can also define a pair correlation function $C(r)$, i.e., the probability that two monomers or sites, separated by a distance r, belong to the same polymer or cluster. For a fractal structure and large r we expect the correlation function to decay as

$$C(r) \sim r^{D_a - d}, \tag{11.21}$$

so that $C(r) \to 0$ as $r \to \infty$. A similar equation also holds for percolation networks at p_c if we replace D_a by D_c, the fractal dimension of the largest percolation cluster. The Fourier transform of $C(r)$ is proportional to the scattered intensity $I(q_s)$ in a static or neutron scattering experiment with a polymer solution, where q_s is the momentum transfer given by

$$q_s = \frac{4\pi}{\lambda} \sin\left(\frac{\theta_s}{2}\right), \tag{11.22}$$

where λ is the wavelength of the radiation, and θ_s is the scattering angle. By Fourier transforming (11.21), it is not difficult to show that

$$I(q_s) \sim q_s^{-D_a}. \tag{11.23}$$

Experimental evidence for (11.20) is actually provided through (11.23). Bouchaud *et al.* (1986) carried out small-angle neutron scattering experiments on a monodisperse polyurethane sample and measured the scattered intensity as a function of q_s. Figure 11.4 presents their results, from which we obtain

$$D_a = 1.98 \pm 0.03, \tag{11.24}$$

in excellent agreement with (11.20). In real applications, polymer solutions are almost always polydisperse and contain polymers of all sizes with radii smaller or larger than ξ. We need to define *average* properties, where the averaging is taken over the polymer size distribution. An average polymer radius is defined by, $\langle R_a \rangle = \sum_{n_a} n_a^2 R_a Q / \sum_{n_a} n_a^2 Q$. Using (11.14) and (11.18) then yields a relation between n_a and $\langle R_a \rangle$, $n_a \sim \langle R_a \rangle^{D_e}$, where D_e is interpreted as the *effective* fractal dimension of *all* branched polymers in the

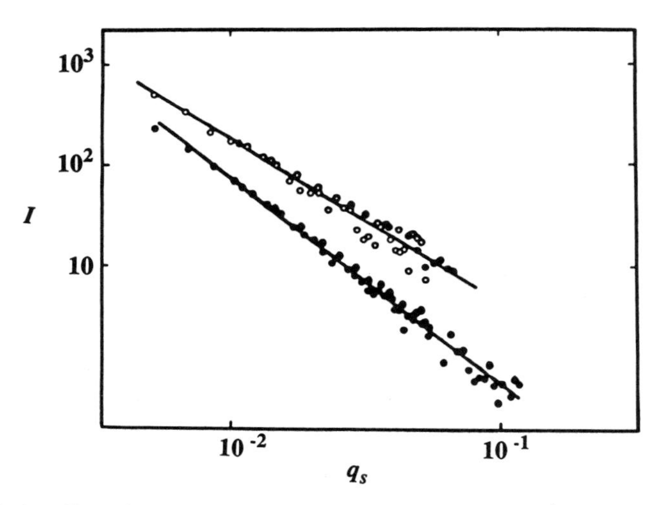

Figure 11.4 *Small-angle neutron scattering results for branched polymers. The upper curve is for a polydisperse polymer solution, whose slope is 1.6. The lower curve is for a single polymer in a dilute good solvent, whose slope is 1.98 (after Bouchaud et al. 1986).*

solution. Daoud *et al.* (1984) showed that for a polydisperse polymer in a dilute solution of a good solvent

$$D_e = D_a(3 - \tau) \tag{11.25}$$

Thus, (11.25) mixes the animal exponent D_a with the percolation or polymer exponent τ. If a percolation description of polymerization is applicable, we should have (see Chapter 2) $\tau(d = 3) = \tau_p(d = 3) \simeq 2.18$, which means that

$$D_e \simeq 1.64, \quad d = 3, \tag{11.26}$$

indicating that the effective fractal dimension is smaller than for a single branched polymer. Since we have defined an effective fractal dimension for a dilute polydisperse polymer solution, the scattering intensity for a solution of polydisperse polymer should also be modified to

$$I(q_s) \sim q_s^{-D_a(3-\tau)}. \tag{11.27}$$

In practice, (11.27) is used in a scattering experiment for estimating D_e and confirming or rejecting the prediction (11.26); this has been done by several sets of careful experiments. For example, Bouchaud *et al.* (1986) synthesized a natural polydisperse polyurethane sample and carried out small-angle neutron scattering on a dilute solution of it. Figure 11.4 shows their results, from which we obtain

$$D_e \simeq 1.6 \pm 0.05, \tag{11.28}$$

in good agreement with (11.26). Adam *et al.* (1987) carried out static light scattering experiments in dilute polydisperse polyurethane solutions and reported $D_e \simeq 1.62 \pm 0.08$, again in good agreement with (11.26). Leibler and Schosseler (1985) measured the average radius of polystyrene, cross-linked by irradiation by elastic light scattering and found $D_e \simeq 1.72 \pm 0.09$, consistent with (11.26). Patton *et al.* (1989) performed both quasielastic and elastic light scattering experiments on branched polyesters and reported $D_e \simeq 1.52 \pm 0.1$, somewhat lower than (11.26) but still consistent with it. A different experiment was carried out by Dubois and Cabane (1989) on a *silica gel*, a physical gel which has a more complex structure than the polymers used by the other researchers (see Chapter 10). Silica gels have a maximum in their scattered intensity curve in the semidilute range, not seen in the other branched polymers discussed above. Moreover, above GP and depending on the pH of the solution, their fractal dimension indicates an anomalous dependence on the concentration, this has not been explained yet. Despite these significant differences, Dubois and Cabane reported that $D_e \simeq 1.58$, quite close to (11.26).

We now compare the experimental data for the scaling behavior of other structural properties of branched polymers and gels and compare them with the percolation predictions. The experimental results for D_e and D_a discussed above all imply that $\tau \simeq 2.2$, in excellent agreement with the percolation prediction, $\tau_p \simeq 2.18$. In their experiments with irradiated polystyrene solution in cyclopentane, Leibler and Schosseler (1985) coupled

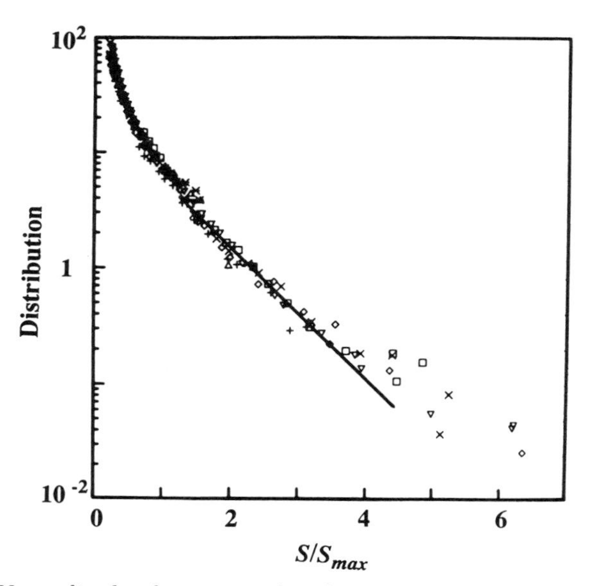

Figure 11.5 *Normalized polymer size distribution as a function of the size s of the polymers, where s_{max} is the maximum size. Percolation predicts that the slope of the line should be* $1 - \tau$, *which then yields* $\tau \simeq 2.3 \pm 0.1$ *(after Leibler and Schosseler 1985).*

gel permeation chromatography and light scattering and obtained the polymer size distribution, which provides a direct means of measuring the exponent τ. Figure 11.5 presents their measurements, from which we obtain $\tau \simeq 2.3 \pm 0.1$, consistent with the percolation prediction, $\tau_p \simeq 2.18$. Lapp *et al.* (1989) further checked this result by doing similar experiments in a system made by chemical end-linking of polydimethylsiloxane, and Patton *et al.* (1989) did the same in a system in which polyester was made by bulk condensation polymerization. The results of both of these groups were consistent with the percolation prediction for the exponent τ. The polymer fractal dimension D_p is related to τ by

$$D_p = \frac{d}{\tau - 1}, \tag{11.29}$$

so these results also imply that $D_p \simeq 2.5$, in agreement with the percolation prediction, $D_c \simeq 2.52$.

Equation (11.15) was tested by Adam *et al.* (1987), who performed static light scattering measurements on a polyurethane sol, and by Candau *et al.* (1985) who carried out their experiments on polystyrene systems cross-linked with divinylbenzene. Figure 11.6 shows the results of Adam *et al.* (1987), from which we obtain $\gamma \simeq 1.71 \pm 0.06$, close to the percolation prediction (see Chapter 2) $\gamma_p \simeq 1.82$. A similar result was obtained by

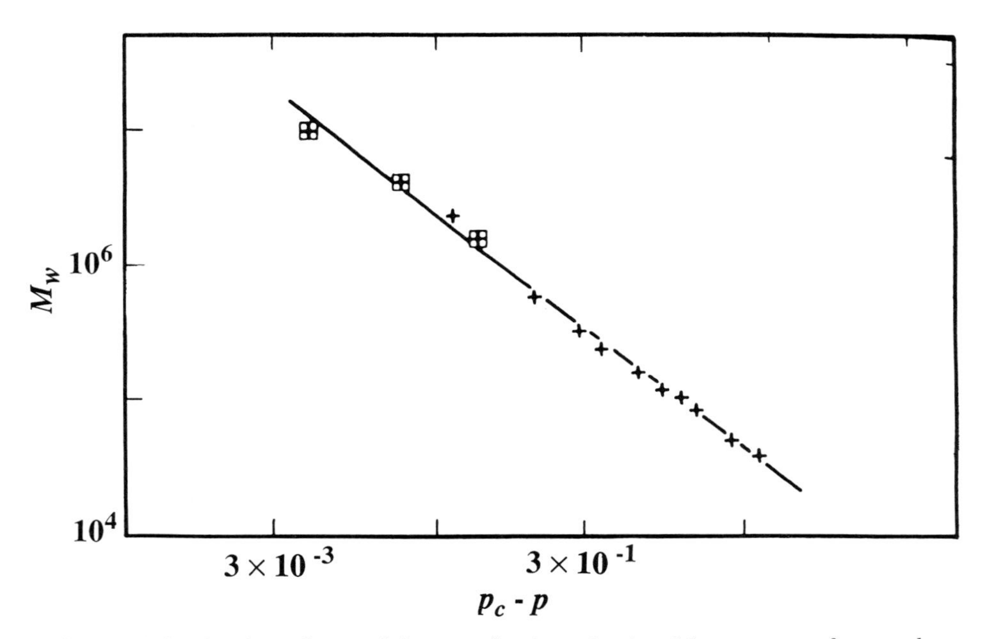

Figure 11.6 *The dependence of degree of polymerization M_w on $p_c - p$ for a polyurethane sol. The slope of the line is $-\gamma$ (after Adam et al. 1987).*

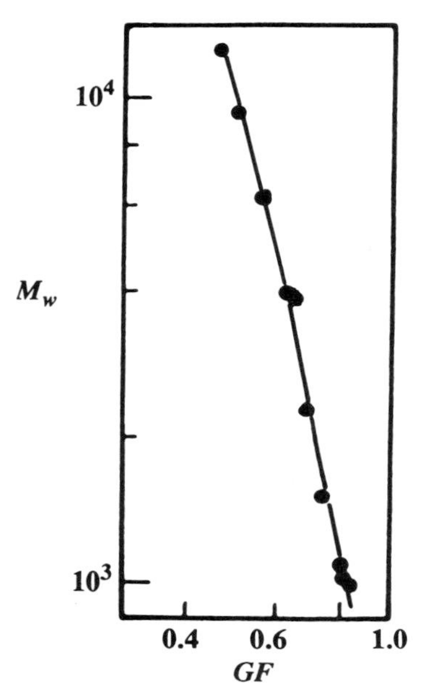

Figure 11.7 *The dependence of the degree of polymerization M_w on the gel fraction GF during anionic copolymerization of divinylbenzene with styrene. The slope of the line is $-\gamma/\beta$ (after Schmidt and Burchard 1981).*

Candau *et al.* (1985). On the other hand, (11.11) and (11.15) can be combined to give $M_w \sim GF^{-\gamma/\beta}$, and thus a plot of $\log(M_w)$ versus $\log(GF)$ can yield an estimate of γ/β. Schmidt and Burchard (1981) carried out anionic copolymerization of divinylbenzene with styrene and obtained branched polymers and gels. Light scattering was used to measure various properties of interest. When they plotted $\log(M_w)$ versus $\log(GF)$, as shown in Fig. 11.7, they obtained a straight line with a slope $\gamma/\beta \simeq 4.5$, in good agreement with the percolation prediction, $\gamma_p/\beta_p \simeq 4.44$. For more recent and accurate experiments regarding these exponents see Trappe *et al.* (1992). We may conclude that three-dimensional percolation provides a very good description of the universal properties of branched polymers and gel networks. Finally, we should point out that the Flory–Stockmayer theory of polymerization predicts that $D_a = D_c = 4$, $\gamma = \beta = 1$, and $\tau = 5/2$, which are nothing but the mean field critical exponents of percolation discussed in Chapter 2, and do not agree with most of the experimental data discussed above.

11.4 Elastic percolation networks

We now describe dynamical properties of percolation networks and their application to modeling of the viscosity and elastic moduli of sols and gels.

Suppose that in a percolation network a fraction p of bonds represent resistors with conductance a and the remaining bonds are resistors with conductance b. Both a and b can be selected from any probability density function $h(g)$ but we consider the simplest case

$$h(g) = p\delta(g - a) + (1 - p)\delta(g - b). \tag{11.30}$$

If $b = 0$ and a is finite, we have a percolation network of conductors and insulators whose conductivity g_e should vanish at $p = p_c$. Near p_c, g_e obeys (2.8). On the other hand, if b is finite and $a = \infty$, we have a percolation network of normal conductors and superconductors whose conductivity g_s is finite for $p < p_c$, but diverges at p_c according to (2.12).

In an analogous way the elastic properties of percolation networks can be defined. Consider a percolation network whose bonds represent elastic springs that can be stretched and/or bent. The elastic energy of the system is given by (Kantor and Webman 1984)

$$E = \frac{\alpha_1}{2} \sum_{\langle ij \rangle} [(\mathbf{u}_i - \mathbf{u}_j) \cdot \mathbf{R}_{ij}]^2 e_{ij} + \frac{\alpha_2}{2} \sum_{\langle jik \rangle} (\delta\boldsymbol{\theta}_{jik})^2 e_{ij} e_{ik}, \tag{11.31}$$

where the first term of the right side represents the contribution of the stretching or central forces (CFs), whereas the second term represents the contribution of angle-changing or bond-bending (BB) forces. Here α_1 and α_2 are the central and BB force constants, respectively, $\mathbf{u}_i = (u_{ix}, u_{iy}, u_{iz})$ is the (infinitesimal) displacement of site i, $\mathbf{R}_{ij}$ is a unit vector from i to j, e_{ij} is the elastic constant of the (spring) between i and j, and $\langle jik \rangle$ indicates that the sum is over all triplets in which the bonds j–i and i–k form an angle whose vertex is at i. The change of angle $\delta\boldsymbol{\theta}_{jik}$ is given by (see also Chapter 4)

$$\delta\boldsymbol{\theta}_{jik} = \begin{cases} (\mathbf{u}_{ij} \times \mathbf{R}_{ij} - \mathbf{u}_{ik} \times \mathbf{R}_{ik}) \cdot (\mathbf{R}_{ij} \times \mathbf{R}_{ik})/|\mathbf{R}_{ij} \times \mathbf{R}_{ik}|, & \mathbf{R}_{ij} \text{ not parallel to } \mathbf{R}_{ik} \\ |(\mathbf{u}_{ij} + \mathbf{u}_{ik}) \times \mathbf{R}_{ij}|, & \mathbf{R}_{ij} \text{ parallel to } \mathbf{R}_{ik} \end{cases} \tag{11.32}$$

where, $\mathbf{u}_{ij} = \mathbf{u}_i - \mathbf{u}_j$. We can now define the elastic properties of percolation networks. Suppose that the elastic constants e_{ij} can be chosen from a distribution like (11.30), i.e., $h(e) = p\delta(e - a) + (1 - p)\delta(e - b)$, i.e., a fraction p of the springs have an elastic constant a and the rest have an elastic constant b. Thus, similar to the case of the electrical conductivity, we can define G_e as the effective elastic moduli of the network when a is finite and $b = 0$, and G_s as the effective elastic moduli of a *superelastic* percolation network when $a = \infty$ (i.e., infinitely rigid bonds) and b is finite. For an experimental realization of a superelastic percolation system see Benguigui and Ron (1993). Then, as discussed in Chapter 2, G_e vanish near p_c with a critical exponent f, whereas G_s diverge as p_c is approached from below with another critical exponent ζ. Unlike the electrical conductivity case, the critical exponents f and ζ are not completely universal; they depend on the presence or absence of BB forces, at least for three-dimensional systems of interest to us. If $\alpha_2 = 0$, only stretching forces are present and we refer to such systems as CF networks. The bond percolation threshold p_{ce} of a CF network is very different from p_{cb}, the usual bond percolation threshold defined in Chapter 2 (Feng and Sen 1984). For a triangular network, $p_{ce} \simeq 0.641$ and $p_{cb} = 2\sin(\pi/18) \simeq 0.347$, and for a BCC network, $p_{ce} \simeq 0.737$ and $p_{cb} \simeq 0.1795$. The reason for this anomalous behavior is that many sample-spanning configurations of a CF network are not rigid, in the sense that all their elastic moduli are zero. In general, for a d-dimensional CF network $p_{ce} \simeq 2d/Z$, where Z is the coordination of the network, so that a CF network has nonzero elastic moduli if $Z > 2d$ and $p > p_{ce}$. Thus, a meaningful study of CF networks is restricted to certain lattices, e.g., the triangular and BCC networks. For d-dimensional cubic networks, $p_{ce} = 1$, so we cannot use them to study percolation in CF systems. We denote the critical exponents of CF networks by f_c and ζ_c. On the other hand, the percolation thresholds of the BB models are the same as those of ordinary percolation if each site of the network interacts with at least $d(d-1)/2$ of its nearest-neighbors in d dimensions. We denote the critical exponents of such BB models by f_b and ζ_b. Near the percolation threshold G_e obeys the following scaling law (Feng and Sen 1984; Kantor and Webman 1984)

$$\begin{cases} G_e \sim (p - \varphi_c)^x, \\ x \ = f_c,\ f_b, \\ \varphi_c = p_{ce},\ p_{cb}, \end{cases} \tag{11.33}$$

whereas G_s follows the power law (Sahimi and Goddard 1985; Feng 1985)

$$\begin{cases} G_s \sim (\varphi_c - p)^{-y}, \\ y \ = \zeta_c,\ \zeta_b. \end{cases} \tag{11.34}$$

Let us describe briefly how the elastic moduli of an elastic or a superelastic percolation network are calculated. We minimize the elastic energy H_E with respect to the nodal displacement $\mathbf{u}_i$, $\partial E/\partial \mathbf{u}_i = 0$. Writing down this equation for every interior node of the network results in a set of dN_n simultaneous equations for a d-dimensional network of N_n internal nodes. The boundary conditions depend on the quantity that we would like to calculate. For example, to calculate the shear modulus of the system, we shear the network by a given strain S_b and impose periodic boundary conditions in the other directions. The resulting set of equations are then solved numerically by, e.g., Gaussian elimination or by an iterative method. From the solution of the set of equations for nodal displacements we

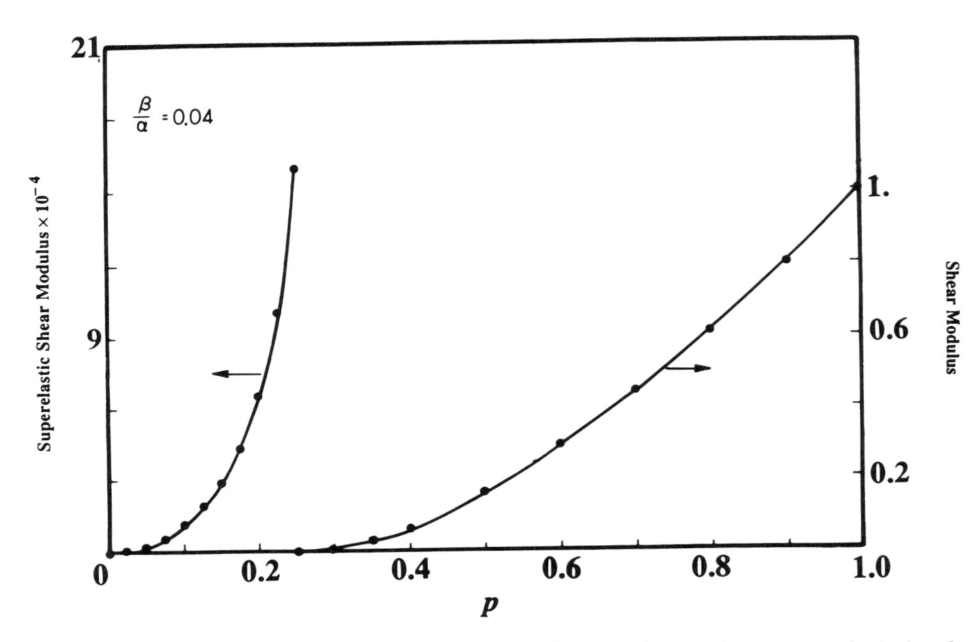

Figure 11.8 *Variations of the shear modulus in elastic and superelastic networks in bond percolation on a simple cubic network. Below the percolation threshold* $p_c \simeq 0.2488$, *p represents the fraction of completely rigid bonds; above* p_c, *p represents the fraction of intact (uncut) bonds.*

calculate E and hence the elastic modulus $G = 2E/S_b^2$. Figure 11.8 shows typical curves for G_e and G_s for a percolation network (compare this figure with Fig. 11.1). The critical exponents are calculated by finite-size scaling discussed in Chapter 2.

Table 11.1 lists the most accurate estimates of these exponents for $d = 3$, along with those in the mean field approximation obtained on a Bethe lattice. Sahimi (1986), and later Roux (1986), proposed that

$$f_b = \mu + 2\nu_p, \tag{11.35}$$

where ν_p is the critical exponent of percolation correlation length. The main idea behind (11.35) is that the scaling of G_e and g_e differ only by a factor $(\text{length})^2$ or ξ_p^2. The predictions of (11.35) seem to be in perfect agreement with the numerical data (see Table 11.1). Unlike f_c and f_b, which seem to be quite different for three-dimensional networks, the numerical values of ζ_c and ζ_b are *not* very different. This is because superelastic networks are dominated by *singly disconnected bonds*, i.e., soft bonds (for which e_{ij} is finite) that connect superelastic (rigid) clusters. If singly a disconnected bond is removed, the two rigid clusters become separated. This is in contrast with elastic percolation networks in which the deformation of blobs (clusters of multiply connected bonds) and the *singly connected bonds*, which connect the blobs, are both important (see Chapter 2). For this reason, we use ζ to denote the critical exponent of G_s for both CF and BB models, and do not distinguish between ζ_b and ζ_c.

Table 11.1 *Critical exponents of elastic and superelastic percolation networks for three-dimensional systems and in the mean field approximation valid for $d \geqslant 6$.*

	μ	s	f_c	f_b	ζ	ζ_d
$d = 3$	2.0	0.73	2.1	3.75	0.65	1.35
$d \geqslant 6$	3	0	3	4	0	0

11.5 Comparison of the scaling properties of the viscosity and elasticity of branched polymers and gels with the percolation predictions

Let us now summarize the experimental data for f, k, and Δ. Most of the experimental measurements of f have been done for what we called chemical gels and have yielded a value of f mostly in the range of 1.9–2.2. Examples include measurements of f for hydrolyzed polyacrylamide (Allain and Salomé 1987), and for tetraethylorthosilicate reactions (Hodgson and Amis 1990). A few measurements for chemical gels have yielded much higher values of f. An example is the measurements of Adam *et al.* (1985) who reported $f \simeq 3.2 \pm 0.6$ for polycondensation. However, all measurements of f for physical gels have yielded $f \simeq 3.8$.

The experimental data for k can be divided into two groups. The first group of measurements indicate a value of k mostly in the range 0.6–0.9. Examples include the work of Adam *et al.* (1979) for copolymerization of styrene divinylbenzene in the reaction bath, and that of Allain and Salomé (1987) for hydrolyzed polyacrylamide. The second group of measurements have mostly yielded a value of k in the range 1.3–1.5. Martin *et al.* (1988) obtained $k \simeq 1.4$ for the viscosity of epoxy resins. Finally, measurements of Δ have all indicated a value around 0.7. Examples are the works of Martin *et al.* (1988), Rubinstein *et al.* (1989), and Hodgson and Amis (1990).

What is the theoretical explanation for the measured values of f and k? de Gennes (1976) proposed that the elastic moduli G of gel polymers near GP and the *conductivity* g_e of percolation networks near p_c belong to the same universality class, that is, $f = \mu$. But it is not clear why the scaling of a *vector* property such as G should be similar to that of a *scalar* property such as g_e. Moreover, even if this analogy were exact, we would still be left with the task of explaining the observed scaling behavior of gel networks because various measurements indicate two distinct values for f, one in the range 1.9–2.2, the other in the range 3.2–3.8; μ is unique for three-dimensional percolation networks. When de Gennes made his suggestion, elastic percolation networks had not been investigated and his suggestion provided a theoretical framework for a qualitative explanation of some of the data.

de Gennes (1979) also suggested that the divergence of the viscosity η of a sol is similar to the divergence of g_s, thus $k = s$. I contend there is *no* analogy between η and g_s because, in general, the rotation and deformation of finite polymers in local viscous shears should make k smaller than s. These deformation and rotational motions also make the field equations for g_s totally different from those for η. Indeed, η is related to a *tensor* quantity, namely, the complex modulus G^*, whereas g_s is a *scalar* quantity, and therefore there is no reason to believe that the scaling of η may be related to that of g_s. Moreover, as in the case of G, even if this analogy were exact, it still could not provide an explanation for the fact that the exponent s is unique for three-dimensional superconducting percolation networks, whereas experimental data for k clearly indicate two distinct values for k. Despite such inconsistencies, these analogies are still used and advocated by some researchers in this field; see the review of Daoud and Lapp (1990).

We now use elastic and superelastic percolation networks and provide a consistent explanation for *all* of the data. In physical gels, BB forces are important. When deformed, any three mutually touching particles roll on top of one another. This motion and the displacement of their centers with respect to one another, create forces that are similar to BB forces. Experimental measurements of f for physical gels (Gauthier-Manuel *et al.* 1988; Woignier *et al.* 1988) confirm this; f is found to be about 3.8, in excellent agreement with the result for three-dimensional BB models, $f_b \simeq 3.75$, estimated by Arbabi and Sahimi (1988). Thus, for physical gels we have

$$f = f_b, \qquad \text{physical gels.} \tag{11.36}$$

In most chemical gels, BB forces are usually not important; the only important force between the monomers seems to be the central or stretching forces. Numerous experimental measurements of f for such gels have yielded a value of f in the range 1.9–2.2, which does not agree with the exponent of the BB model but agrees nicely with the exponent of the three-dimensional CF models, $f_c \simeq 2.1$, estimated by Arbabi and Sahimi (1993). Therefore, in this case we have

$$f = f_c, \qquad \text{chemical gels.} \tag{11.37}$$

The fact that the critical exponent f_c is close to the conductivity exponent μ does not imply that the two models belong to the same universality class. This scaling picture is perfectly consistent with what we discussed above and with what we know about physical and chemical gels. Alexander (1984) argued that in some gels and rubbers under *internal or external* stress, some terms in the elastic energy of the system are similar to the so-called Born model, the elastic energy of which is given by

$$E = \frac{\alpha_1}{2}\sum_{\langle ij\rangle}[\mathbf{u}_i - \mathbf{u}_j)\cdot\mathbf{R}_{ij}]^2 e_{ij} + \frac{\alpha_3}{2}\sum_{\langle ij\rangle}(\mathbf{u}_i - \mathbf{u}_j)^2 e_{ij}, \qquad (11.38)$$

where α_3 is another constant. In three dimensions the critical exponent of this system is $f_B = \mu \simeq 2$. The second term of the right side of (11.38) is purely a scalar term and is similar to (11.30). Near p_c the second term dominates the first term (which is due to central forces). But these rubbers and gels differ from the Born model in several important ways, such as the presence of nonlinear terms in their elastic energy and the possibility of negative as well as positive Born coefficients. It is not clear, therefore, that the Alexander result $f = f_B = \mu$ is applicable to such systems. With our interpretation of the experimental data, in the polymers used by Adam *et al.* (1985) the BB forces might be important since they reported that, $f \simeq 3.2 \pm 0.6$, consistent with (11.36). In some polymers entropic effects are important, whereas in the elastic energy given by (11.31) such effects cannot be taken into account. It is not yet clear that if such effects are taken into account, the various elasticity critical exponents would remain the same as those found with (11.31). Daoud and Coniglio (1981) suggested that the elasticity exponent for such systems is equal to $\nu_p d$. For three-dimensional systems this predicts $f \simeq 2.64$, quite different from the exponents of CF and BB models. Whether the Daoud–Coniglio relation is exact and how the crossovers between these various regimes take place are still not understood, but the result of Adam *et al.* (1985), $f \simeq 3.2 \pm 0.6$, can also be interpreted with the Daoud–Coniglio relation. Thus, we may have *three* universality classes for the critical properties of the elasticity of polymers, but this possibility remains an open question.

How can we explain the experimental data on the scaling behavior of η near the GP? Sahimi and Goddard (1985) proposed that the scaling behavior of η near the GP is analogous to that of the shear modulus G_s of a superelastic percolation network near p_c. The *relation* between η and the shear modulus of an appropriate elastic system can be inferred, in a straightforward and *rigorous* way, from the continuum equations of elasticity, see, e.g., Christensen (1982). The viscosity coefficient that is obtained from such a continuum theory of elasticity is actually the zero-shear rate viscosity, for it is necessary to have a vanishingly small rate of deformation in order to be able to define the viscosity coefficient. The *analogy* between η and the shear modulus G_s of a superelastic percolation network is made simply because they both diverge as the percolation threshold, or GP, is approached. Even the analogy between η and G_s is not nearly enough to explain the scaling of η near the GP since, as reviewed above, experimental data indicate that the value of k is either in the range 0.6–0.9, or in the range 1.3–1.5, whereas the shear modulus G_s of a three-dimensional superelastic percolation network is characterized by a *unique* value of ζ. As Arbabi and Sahimi (1990) explained (see also Sahimi and Arbabi 1993), the reason for having two distinct values of k is that the dynamics of the two systems are totally different. In one case, the dynamics of the solution corresponds to the so-called Zimm limit, while in the other case the solution obeys the so-called Rouse dynamics.

In the Zimm limit there are strong hydrodynamic interactions between monomers and also between polymers of various sizes. In this regime, polymeric cluster relaxation is governed by Stokes–Einstein diffusion. The surroundings of any polymer are composed of smaller polymers which are stationary on the molecular time scale. Thus, there is no significant diffusion of the polymers in the reaction bath. Hence, a superelastic percolation network, which is a *static* system in which the totally rigid clusters are motionless may be used to simulate the Zimm regime. Indeed, the estimated value of ζ for such a system, $\zeta \simeq 0.65$, obtained by Arbabi and Sahimi (1990) supports the idea that gelling solutions with k values 0.6–0.9 are in fact close to the Zimm regime. Moreover, Arbabi and Sahimi (1990) proposed that

$$k = \zeta = \nu_p - \frac{\beta_p}{2}, \qquad \text{Zimm regime.} \tag{11.39}$$

Kertész (1983) and Sahimi and Hughes (see Sahimi 1983) had suggested that $s = \nu_p - \beta_p/2$, using various arguments. However, it is now clear that $s > \nu_p - \beta_p/2$, (except for $d \geqslant 6$, where $s = \nu_p - \beta_p/2 = 0$), and that (11.39) may be exact. This is also consistent with the suggestion of Limat (1989) that $\zeta < s$, and with the experimental result of Benguigui and Ron (1993), $\zeta \simeq 0.67$.

On the other hand, the gelling solution can also be in a state in which there are no hydrodynamic interactions between the polymers of various sizes (i.e., the Rouse regime), and therefore the polymers can diffuse essentially freely in the reaction bath. Hence, a static superelastic percolation network cannot model this regime of polymerization. To model this regime Arbabi and Sahimi (1990) proposed a *dynamic* superelastic network in which each cluster of the rigid bonds represents a polymer, and there is of course a wide size distribution of such polymers or clusters in the network. The soft bonds (those for which the elastic constant e_{ij} is finite) represent the liquid (solution) medium in which the rigid clusters move randomly, with equal probability, in one principal direction of the network. This simulates the diffusion of the polymers in the reaction bath. Two rigid clusters cannot overlap but they can temporarily join and form a larger cluster, which can be broken up again at a later time. It was shown that the shear modulus of this dynamic system diverges with an exponent ζ_d which, in three dimensions, is $\zeta_d \simeq 1.35$. This supports the idea that those gelling solutions whose values of k are in the range 1.3–1.5 are in fact close to the Rouse limit. Moreover, Arbabi and Sahimi (1990) proposed that

$$k = \zeta_d = 2\nu_p - \beta_p, \qquad \text{Rouse limit.} \tag{11.40}$$

The relation, $k = 2\nu_p - \beta_p$ had already been suggested by de Gennes (1979). What Arbabi and Sahimi (1990) did was to show that the elastic moduli of a *dynamic* superelastic percolation network have the same critical exponent. Therefore, static and dynamic elastic and superelastic percolation networks provide a consistent explanation for the scaling properties of the viscosity of gelling solutions below and near GP, and those of the elastic moduli of the gel network above and close to GP. The Flory–Stockmayer theory predicts that $f = 3$, and that η diverges as $\eta \sim \log(p_c - p)$, i.e., $k = 0$, in disagreement with the experimental data.

There are some experimentally observed deviations of k from ζ or ζ_d. As reviewed above, experimental determination of k and f usually involves measuring the complex shear modulus G^* at a finite (albeit small) frequency ω. Strictly speaking, the scaling laws for G and η are valid only in the limit $\omega \to 0$, whereas in practice it is highly difficult to achieve such a limit, therefore the reported values of k show some deviations from ζ or ζ_d. These deviations can be attributed to transient effects, which should diminish as lower frequencies are achieved. Using this scaling description for G and η, we can find upper and lower bounds for the exponent $\Delta = f/(f + k)$. If for chemical gels we use $f = f_c \simeq 2.1$ and use (11.39) and (11.40) for estimating k, we obtain

$$0.61 \leqslant \Delta \leqslant 0.75, \qquad (11.41)$$

where the lower and upper limits correspond to the Rouse and Zimm regimes, respectively. Thus, changing k by a factor of two causes only a 20% change in Δ. Moreover, the *average* value of Δ is about 0.68, very close to the measured value $\Delta \simeq 0.7$. Thus, because of the relative insensitivity of Δ to the value of k, measurement of Δ is *not* an accurate means of assessing the dynamics of gelling solutions.

11.6 Kinetic gelation and dynamic percolation

The large elastic moduli and low solvent absorption of some highly cross-linked polymers make them highly desirable for various applications. They can be made by free-radical homopolymerization of tetrafunctional monomers such as diacrylates and dimethacrylates, which polymerize very rapidly at room temperature if initiated by ultraviolet light and suitable photoinitiator. Typical results of such a homopolymerization include auto-acceleration kinetics, formation of a glassy polymer, and attainment of a maximum conversion beyond which no reaction occurs even when un-reacted initiator is still in the reactor. The percolation model discussed above cannot be used for free-radical polymerization, since this is a kinetic process and random percolation is a static phenomenon. Although several altern-ative models have been proposed, we describe kinetic gelation, a variation of the percolation model that appears to explain many features of free-radical polymerization.

Kinetic gelation was first proposed by Manneville and de Seze (1981). In their model each site of a lattice represents either a di- or tetrafunctional monomer. At time $t = 0$ a fraction of the sites are assumed to represent an active radical. This is called *fast initiation*, since it is done before any polymerization has taken place. At each time step an active radical is selected at random and moved to an adjacent site that has at least one unreacted functional group. A bond is formed between the two sites and the new site also becomes a radical. If a site is occupied simultaneously by two radicals, then termination occurs. Radicals in an actual free-radical polymerization sometimes have greatly reduced mobility, and can have lifetimes of a year or even longer. This phenomenon is also observed in the kinetic gelation model of Manneville and de Seze when none of the nearest-neighbor sites of a radical has unreacted functional units; it is called *radical trapping*. In any event, it is possible to form a sample-spanning gel similar to a percolation cluster. As such, kinetic gelation can be thought of as a dynamic percolation process, different from random and static percolation discussed so far.

Herrmann *et al.* (1982) and Bansil *et al.* (1984) modified this basic model by adding the effect of a solvent and a mobile monomer. This does not affect the GP but decreases slightly the effect of radical trapping. Certain trends observed in experiments, such as an increase in the GP conversion either as the fraction of tetrafunctional monomers decreased, or as initiator concentration increased, or as solvent concentration increased, were reproduced by this model. Boots *et al.* (1985) developed a kinetic gelation model with *slow initiation* in which only one active radical is present in the lattice at any given time. At time $t = 0$ a monomer is randomly selected and designated as an active radical. When this radical is surrounded by sites without unreacted functional groups and is trapped, another randomly selected monomer is initiated. Therefore, no termination of radicals can occur in this model. It can adequately predict the pendant double-bond fraction per monomeric unit in the polymer for the reaction of 1,6-hexanediol diacrylate, and can give rise to inhomogeneous structures, which result in higher reactivity at low conversions of the pendant double bond relative to the monomeric double bond. Such inhomogeneous structures have been obtained in various experiments.

Simon *et al.* (1989) have used kinetic gelation models to determine, within experimental error, the fraction of constrained and unconstrained double bonds over a wide range of conversions in the polymerization of ethylene glycol dimethacrylate. The rules for determining constraint in the model are that all pendant double bonds and all monomers in a pool of six or less are constrained. Monomers in pools of seven or more are assumed to be unconstrained. The reactivity is not affected by whether a site is constrained or not. The structure of the gel networks formed during kinetic gelation has been analyzed (see, e.g., Herrmann *et al.* 1982), and it has been shown that the critical exponents characterizing it are the same as those of random percolation described above and in Chapter 2. Thus, kinetic gelation models, which are nothing but dynamic percolation processes, appear to provide adequate models of free-radical polymerization. And free-radical polymerization is far more complex than ordinary gelation phenomena. Moreover, many other variants of the random percolation model have been invented for studying certain aspects of polymerization, e.g., for taking into account effects of the solvent which we ignored. An excellent discussion of these models is given by Stauffer *et al.* (1982).

11.7 Conclusions

In this chapter, we discussed the available experimental data on structural, mechanical, and rheological properties of branched polymers and gels, and showed that appropriate percolation models may be developed that can provide theoretically consistent explanations for most if not all of the experimental data.

References

Adam, M., Delsanti, M., and Durand, D., 1985, *Macromolecules* **18**, 2285.
Adam, M., Delsanti, M., Munch, J.P., and Durand, D., 1987, *J. Physique* **48**, 1809.
Adam, M., Delsanti, M., Okesha, R., and Hild, G., 1979, *J. Physique Lett.* **40**, L539.
Alexander, S., 1984, *J. Physique* **45**, 1939.
Allain, C. and Salomé, L., 1987, *Polymer Commun.* **28**, 109.
Arbabi, S., and Sahimi, M., 1988, *Phys. Rev. B* **38**, 7173.
Arbabi, S., and Sahimi, M., 1990, *Phys. Rev. Lett.* **65**, 725.
Arbabi, S., and Sahimi, M., 1993, *Phys. Rev. B* **47**, 695.
Bansil, R., Herrmann, H.J., and Stauffer, D., 1984, *Macromolecules* **17**, 998.
Benguigui, L. and Ron, P., 1993, *Phys. Rev. Lett.* **70**, 2423.
Boots, H., Kloosterboer, J., van de Hei, G., and Pandey, R.B., 1985, *Brit. Polymer J.* **17**, 219.
Bouchaud, E., Delsanti, M., Adam, M., Daoud, M., and Durand, D., 1986, *J. Physique Lett.* **47**, 539.
Candau, S.J., Ankrim, M., Munch, J.P., Rempp, P., Hild, G., and Osaka, R., 1985, in *Physical Optics of Dynamical Phenomena in Macromolecular Systems*, Berlin, De Gruyter, p. 145.
Christensen, R.M., 1982, *Theory of Viscoelasticity*, 2nd ed., New York, Academic Press, p. 12.
Cotton, J.P., 1974, *Thése, Université Paris*, 6.
Daoud, M., 1988, *J. Phys. A* **21**, L937.
Daoud, M. and Coniglio, A., 1981, *J. Phys. A* **14**, L301.
Daoud, M., Family, F., and Jannik, G., 1984, *J. Physique Lett.* **45**, 199.
Daoud, M. and Lapp, A., 1990, *J. Phys. Condens. Matter* **2**, 4021.
de Gennes, P.G., 1971, *Phys. Lett.* **A38**, 339.
de Gennes, P.G., 1976, *J. Physique Lett.* **37**, L1.

de Gennes, P.G., 1979, *J. Physique Lett.* **40**, L197.
Dubois, M. and Cabane, B., 1989, *Macromolecules* **22**, 2526.
Durand, D., Delsanti, M., Adam, M., and Luck, J.M., 1987, *Europhys. Lett.* **3**, 297.
Family, F. and Coniglio, A., 1980, *J. Phys. A* **13**, L403.
Feng, S., 1985, *Phys. Rev. B* **32**, 510.
Feng, S. and Sen, P.N., 1984, *Phys. Rev. Lett.* **52**, 216.
Ferry, J.D., 1980, *Viscoelastic Properties of Polymers*, New York, Wiley.
Flory, P.J., 1941, *J. Am. Chem. Soc.* **63**, 3083.
Gauthier-Manuel, B., Guyon, E., Roux, S., Gits., S., and LefauCheux, F., 1988, *J. Physique* **48**, 869.
Herrmann, H.J., Landau, D.P., and Stauffer, D., 1982, *Phys. Rev. Lett.* **49**, 412.
Hodgson, D.F. and Amis, E.J., 1990, *Macromolecules* **23**, 2512.
Holly, E.E., Venkataraman, S.K., Chambon, F., and Winter, H.H., 1988, *J. Non-Newtonian Fluid Mech.* **27**, 17.
Kantor, Y. and Webman, I., 1984, *Phys. Rev. Lett.* **52**, 1891.
Kertész, J., 1983, *J. Phys. A* **16**, L471.
Lapp, A., Leibler, L., Schosseler, F., and Strazielle, C., 1989, *Macromolecules* **22**, 2871.
Leibler, L. and Schosseler, F., 1985, *Phys. Rev. Lett.* **55**, 1110.
Limat, L., 1989, *Phys. Rev. B* **40**, 9253.
Lin, Y.G., Mallin, D.T., Chien, J.C.W., and Winter, H.H., 1991, *Macromolecules* **24**, 850.
Lubensky, T.C. and Isaacson, J., 1978, *Phys. Rev. Lett.* **41**, 829.
Manneville, P. and de Seze, L., 1981, in *Numerical Methods in the Study of Critical Phenomena*, edited by J. Della Dora, J. Demongeot and B. Lacolle, Berlin, Springer, p. 116.
Martin, J.E., Adolf, D., and Wilcoxon, J.P., 1988, *Phys. Rev. Lett.* **61**, 2620.
Parisi, G. and Sourlas, N., 1981, *Phys. Rev. Lett.* **46**, 891.
Patton, E., Wesson, J.A., Rubinstein, M., Wilson, J.E., and Oppenheimer, L.E., 1989, *Macromolecules* **22**, 1946.
Roux, S., 1986, *J. Phys. A* **19**, L351.
Rubinstein, M., Colbey, R.H., and Gillmore, J.R., 1989, *Polym. Prepr. Am. Chem. Soc. Div. Polym. Chem.* **30** (1), 81.
Sahimi, M., 1983, in *The Mathematics and Physics of Disordered Media*, edited by B. D. Hughes and B. W. Ninham, *Lecture Notes in Mathematics 1035*, Berlin, Springer, p. 339.
Sahimi, M., 1986, *J. Phys. C* **19**, L79.
Sahimi, M. and Arbabi, S., 1993, *Phys. Rev. B* **47**, 703.
Sahimi, M. and Goddard, J.D., 1985, *Phys. Rev. B* **32**, 1869.
Schmidt, W. and Burchard, W., 1981, *Macromolecules* **14**, 370.
Simon, G.P., Allen, P.E.M., Bennet, D.J., Williams, D.R.G., and Williams, E.H., 1989, *Macromolecules* **22**, 3555.
Stauffer, D., 1976, *J. Chem. Soc. Faraday Trans.* **II 72**, 1354.
Stauffer, D., Coniglio, A., and Adam, M., 1982, *Adv. Polymer Sci.* **44**, 105.
Stockmayer, W.H., 1943, *J. Chem. Phys.* **11**, 45.
Trappe, V., Richtering, W., and Burchard, W., 1992, *J. Physique II France* **2**, 1453.
Woignier, T., Phalippou, J., Sempere, R., and Pelons, J., 1988, *J. Physique. France* **49**, 89.

12
Morphological and transport properties of composite materials

12.0 Introduction

Composite materials and media constitute a large class of naturally occurring or man-made disordered systems. Included in this class are granular materials, composite solids, metal–insulator films, porous media, polymers and aggregates, colloids, and many others. Currently, composite materials are the subject of intensive research, not only for the fundamental scientific questions that have been raised by their interesting structural and transport properties, but also for their practical applications. In this chapter we discuss the application of percolation concepts to predicting the effective properties of composite media. Calculation of the effective properties of porous media and gel polymers is discussed in earlier chapters, so we do not considered it here. Far from the percolation threshold p_c, the effective properties of composite materials can be accurately predicted by the effective-medium approximation (EMA) discussed in Chapter 5. Garland and Tanner (1978), Lafait and Tanner (1989), and Cody *et al.* (1990) contain references on the applications of Chapter 5 methods to composite media. As in Chapter 11, we restrict most of our discussion to the scaling behavior of the composites in the critical region near p_c, where percolation effects are dominant and universal properties, independent of the microscopic details of the system, may be observed. Later we shall see that the critical region is often quite broad, therefore such universal scaling laws can be very useful for predicting the effective properties of a composite medium over a broad range of the volume fraction of the components constituting the composite.

12.1 DC conductivity and elastic moduli of powders and polymer composites

One of the simplest three-dimensional composite systems is a random close packing of hard spheres. The spheres can be considered as one phase of the material, while the host medium between the spheres, or the matrix, constitutes another phase. Although this two-phase system may seem too simple, its effective properties are directly relevant to those of a wide variety of systems of practical importance. For example, if the spheres are conducting and the matrix is insulating, the packing can be used for studying the electrical conductivity of disordered composites such as powders. If we imagine that the matrix is the pore space through which a fluid can flow, the packing can be used for studying flow phenomena in unconsolidated porous rocks, a subject discussed in Chapters 5 to 7. Finally, rheological properties of a system in which hard spheres are suspended in a fluid have been of interest for a long time. It has been found that the viscosity of the suspension follows the following scaling law

$$\eta \sim (\phi_c - \phi)^{-m}, \tag{12.1}$$

where ϕ is the volume fraction of the solid spheres, and ϕ_c is its critical value at which the viscosity diverges. Equation (12.1) is similar to (11.2) for the viscosity of a gelling solution. Thus, percolation is certainly relevant to modeling the viscosity of a suspension.

Historically, the work of Malliaris and Turner (1971) was probably the first in which the electrical conductivity of a compacted powder of spherical particles was measured, and a clear reference was made to percolation. These authors prepared powder samples of high-density polyethylene particles (with radius $R_p = 150\mu\text{m}$) and nickel particles (with radius $R_n = 4$–$7\,\mu\text{m}$). They then observed that the electrical conductivity of the powder was essentially zero unless "the composition of the metal reached a critical value." This critical value, the percolation threshold of the metallic phase, was found to depend on R_p/R_n. The electrical resistivity of the powder dropped 20 orders of magnitude at the percolation threshold $p_c \simeq 0.06$ for samples with nickel segregated on the outside of the polyethylene spheres, and at $p_c \simeq 0.35$ for samples prepared so that the nickel penetrated the spheres. Malliaris and Turner (1971) also remarked that, "the critical composition for a sudden increase in the electrical conductivity of the system was assumed to correspond to the first nonzero probability for infinitely long chains of contiguously occupied lattice sites," a clear reference to the formation of a percolating cluster of metallic particles. Although they did not refer to percolation explicitly, they used bond percolation thresholds of various lattices computed by Domb and Sykes (1961) to interpret their experimental data.

Fitzpatrick *et al.* (1974) were the first to explicitly study percolation and electrical conduction in random close packed spheres, as part of a first-year physics course at Harvard University. All the spheres had the same diameter. The conducting spheres were made of aluminum, the insulating ones of acrylic plastic. The system was pressed between two parallel electrodes made of crumpled aluminum, and the measurements were made with an ohmmeter at high resistances and with a simple Wheatstone bridge at low resistances. The overall conductivity of the system was measured as a function of the volume fraction of the conducting spheres. It was observed that the conductivity of the system vanishes at a finite value of the volume fraction of the conducting spheres. Independently, Clerc *et al.* (1975) studied mixtures of conducting and insulating confectioner's dragées, which were neither spherical nor closely calibrated.

The first systematic and extensive study of percolation and conduction properties of packings of particles appears to have been carried out by Ottavi *et al.* (1978). They carried out experiments with molded plastic spheres, all having the same diameter, in which a fraction of them were electroplated with a copper coating to make them conducting. The spheres were poured into a cylinder equipped with a pair of electrodes, one on the rigid bottom

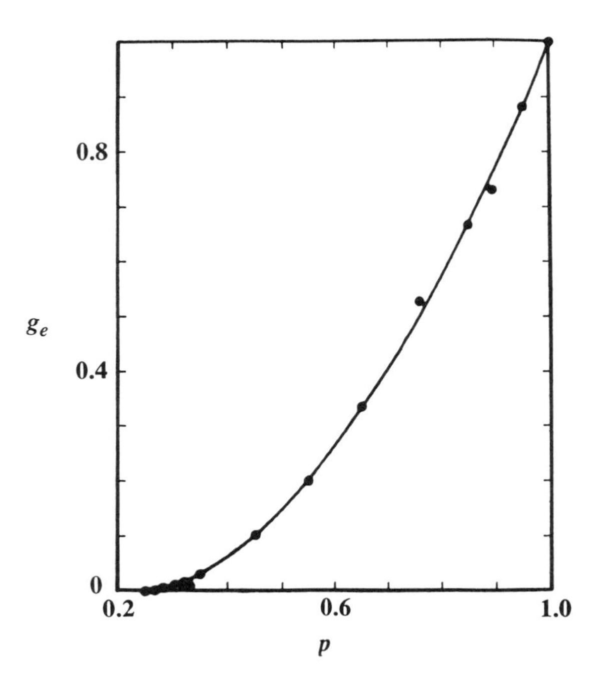

Figure 12.1 *Variations of the electrical conductivity g_e of a mixture of conducting and nonconducting spheres as a function of the fraction p of the conducting spheres (after Ottavi et al. 1978).*

of the cylinder, the other at the top. Their results, shown in Fig. 12.1, indicate that the effective conductivity g_e vanishes at $p_c \simeq 0.29 \pm 0.02$. Since the filling factor f_l (see Chapter 2) of such packings is about 0.6, we obtain $\phi_c = f_l p_c \simeq 0.17$, in agreement with the theoretical prediction of Scher and Zallen (1970) for the percolation threshold of random three-dimensional percolating continua discussed in Chapter 2. The data of Ottavi *et al.* (1978) also indicated that near p_c

$$g_e(p) \sim (p - p_c)^{\mu}, \tag{12.2}$$

with $\mu \simeq 1.7 \pm 0.2$, reasonably close to $\mu \simeq 2.0$ for three-dimensional percolation discussed in Chapter 2. Deptuck *et al.* (1985) investigated sintered, submicrometer silver powder with a volume fraction $\phi \sim 0.4$, commonly used for millikelvin and submillikelvin cryostats. The sinter remains elastic and percolating for ϕ as low as 0.1. They also investigated submicrometer copperoxide–silver powder, routinely used in heat exchangers for optimiz-

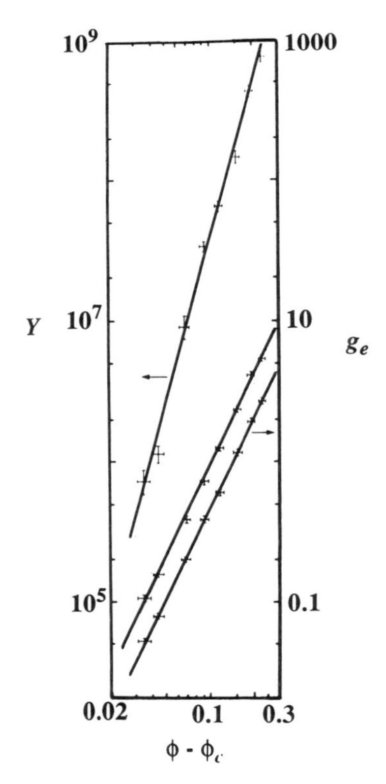

Figure 12.2 *Conductivity and Young's modulus of sintered, submicrometer, silver-powder beams as a function of* $(\phi - \phi_c)$, *where* $\phi_c \simeq 0.062$ *The two conductivity curves represent the data at 78K (upper curve) and 300K (after Deptuck et al. 1985).*

ing heat transfer to dilute liquid ^3He-^4He mixtures. In these composites, the silver component behaves as a percolating system with $\phi_c < 0.1$. Deptuck *et al.* carried out a systematic study of the electrical conductivity and the Young modulus of these powders and measured them in the *same* range of ϕ. This made it possible to compare their critical behavior in the same range of ϕ. Their results, shown in Fig. 12.2, indicate that the conductivity vanishes with a critical exponent $\mu \simeq 2.15 \pm 0.25$, completely consistent with $\mu \simeq 2.0$ for three-dimensional percolation, while the Young modulus G vanishes according to the scaling law

$$G \sim (\phi - \phi_c)^f, \tag{12.3}$$

with $f \simeq 3.8$, in excellent agreement with the critical exponent $f_b \simeq 3.75$ of elastic percolation networks with stretching and bond-bending forces discussed in Chapter 11. As Fig. 12.2 indicates, the critical region is quite broad. Similar results were obtained by Lee *et al.* (1986) who used silver-coated glass–Teflon composites. The Teflon powder used was composed of particles with diameters of about 1 μm. The powder was then mixed with glass spheres coated with a 600Å silver layer which provided high conductivity. The conductivity of the mixture was then measured as a function of the volume fraction of the conducting particles. Lee *et al.* found that $\phi_c \simeq 0.17$, in good agreement with the percolation threshold of three-dimensional continuum percolation predicted by Scher and Zallen (1970). Moreover, when they fitted their conductivity data to (12.2), they found $\mu \simeq 2.0 \pm 0.2$, in excellent agreement with that of three-dimensional percolation conductivity. Lee *et al.* also found that the critical region in which (12.2) was applicable was quite broad. Deprez *et al.* (1989), who measured the electrical conductivity of sintered nickel samples, also found that (12.2) (with $\mu \simeq 2.0$) was applicable over the *entire range* of the volume fraction.

Hsu and Berzins (1985) studied percolation effects in perfluorinated ionomers, with general formula, $[(CF_2)_nCF]_m$–O–R–SO_3X, where n is typically 6–13, R is a perfluoroalkylene group which may contain ether oxygen, and X is any monovalent cation. They are comprised of carbon–fluorine backbone chains with perfluoro side chains containing sulfonate, carboxylate, or sulfonamide groups, possess exceptional transport, chemical, and mechanical properties, and have been used as membrane separators, acid catalysts, and polymer electrodes. Percolation effects are important in these composites because of a spontaneous phase separation occuring in the wet polymer; the conductive aqueous phase is distributed randomly in the insulative fluorocarbon phase. The effective properties of these polymers are dominated by the distribution and connectivity of the clusters of the conductive phase in the midst of an insulating phase; this is a percolation phenomenon. Hsu and Berzins (1985) measured the electrical conductivity and elastic moduli of these polymer composites, and found their measurements to be consistent with the predictions of percolation theory. For example, they found that for $10^{-2} \leqslant p - p_c \leqslant 0.8$, a very broad region, the conductivity of the polymers obeyed (12.2), which again indicates the broad applicability and usefulness of a universal scaling law such as (12.2).

Many composite materials are anisotropic and the effect of the anisotropy on their effective properties near p_c is particularly important. Smith and Lobb (1979) measured the conductivities of two-dimensional conductor–insulator networks generated photolithographically from laser speckle patterns. When they measured the conductivity of isotropic samples, they found that it vanishes at $p_c \simeq 0.59$, in agreement with the site percolation threshold of a square network, with a critical exponent $\mu \simeq 1.3$, in complete agreement with two-dimensional percolation conductivity (see Chapter 2). When they measured the conductivities of anisotropic samples, they found that the conductivity anisotropy, measured in terms of the difference between the conductivity of the system in different directions, decreased as p_c was approached.

Troadec *et al.* (1981) measured thermal and electrical conductivity of conducting WTe_2 and semiconducting WSe_2 powders, characterized both by a geometrical anisotropy (grain shapes and sizes) and by anisotropy in a transport property. These mixtures belong to the family of dichalcognides of transition metals, TX_2. Their structure is layered with a hexagonal arrangement within each plane, which gives rise to the anisotropy of the

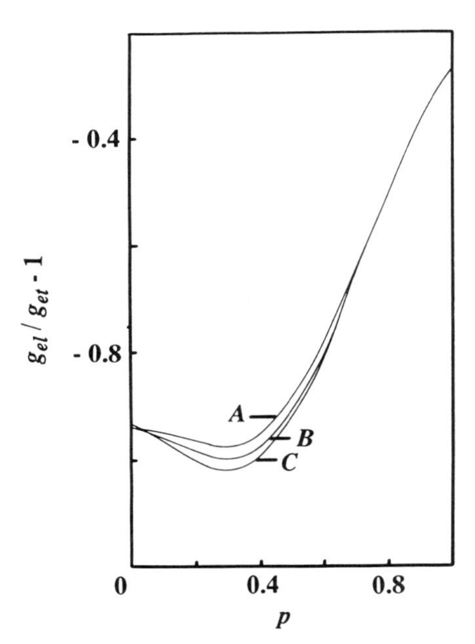

Figure 12.3 *Measured conductivity anisotropy $g_{e\ell}/g_{et} - 1$ of a mixture of WTe_2 and* WSe_2 *powders, as a function of the fraction p of WTe_2 in the mixture, at (A) $T = 300K$, (B) $T = 200K$, and (C) $T = 100K$ (after Troadec et al. 1981).*

system. The main difference between the crystallographic structures of WSe_2 and WTe_2 is in the coordination number of the W atoms. In WSe_2, W is at the center of a trigonal prism of Se atoms, whereas in WTe_2, W has an octahedral environment but is not precisely at the center of the octahedron. Moreover, WSe_2 has a semiconductor character, in contrast with the semimetallic character of WTe_2. Troadec *et al.* used powder grains that were single crystal platelets having a thickness of about 1 to 2 μm and a horizontal hexagonal shape with a largest dimension of about 20 μm. The mixed powder was outgassed under secondary vacuum for 2h before being sintered in the same pressure and temperature conditions. The packing fraction was about 92%. Figure 12.3 shows their measured anisotropy in the electrical conductivity, expressed as $g_{e\ell}/g_{et} - 1$, where $g_{e\ell}(g_{et})$ is the conductivity in the direction parallel (perpendicular) to the applied potential, and p is the fraction of WTe_2 in the mixture. The anisotropy shows a minimum at the percolation threshold of the WTe_2 phase. This is expected, since as the percolation threshold of a phase is approached, its conducting paths become very tortuous, so that the current cannot distinguish between different directions. Indeed, Shklovskii (1978) suggested that near p_c we must have

$$\frac{g_{e\ell}}{g_{et}} - 1 \sim (p - p_c)^{\lambda_p}, \tag{12.4}$$

where λ_p is a universal critical exponent. If the system always remains conducting, then the anisotropy should be minimum at p_c, and this is consistent with the data shown in Fig. 12.3. The conductivity of anisotropic systems can also be predicted by EMA, (5.12)–(5.14), and indeed EMA predicts that the conductivity anisotropy vanishes at p_c. It is not clear whether λ_p is related to the other percolation exponents or whether it is a completely independent exponent.

Results very similar to those of Troadec *et al.* were also obtained by Balberg *et al.* (1983), who measured the resistivity of a composite composed of elongated carbon black aggregates embedded in an insulating plastic, polyvinylchloride. Because of the elongation of the aggregates, the system is anisotropic. Measurements of the anisotropy of the system produced results very similar to those shown in Fig. 12.3, except that the curves ended at the percolation threshold, since the plastic matrix was insulating, in agreement with (12.4).

What is the effect of the thickness of the sample on the percolation properties of such systems? It is not difficult to show that

$$p_c(\ell) - p_c(\infty) \sim c\left(\frac{R_p}{\ell}\right)^{1/\nu_{p^3}}, \tag{12.5}$$

where $p_c(\ell)$ is the effective percolation threshold of a sample of thickness ℓ, $p_c(\infty)$ is the corresponding threshold for $\ell \to \infty$, c is a constant, and ν_{p^3}

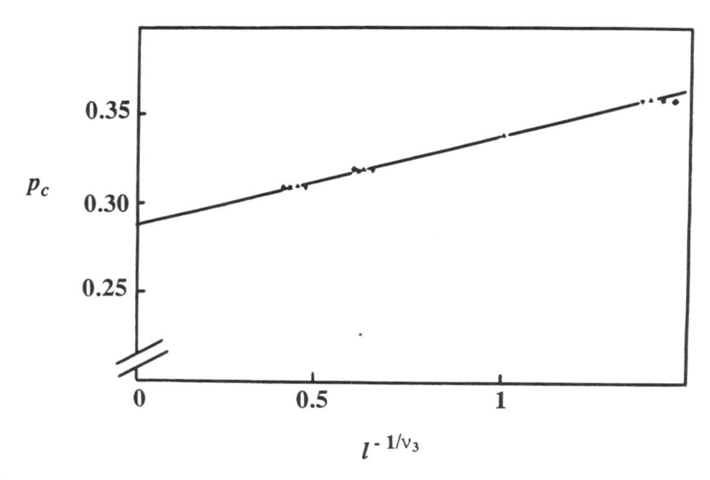

Figure 12.4 *Variations of the percolation threshold p_c of a mixture of conducting and insulating spheres with the thickness l of the sample. ν_3 is the correlation length exponent of three-dimensional percolation (after Clerc et al. 1980).*

is the correlation length for three-dimensional percolation. Equation (12.5) is identical to (2.20), where we discussed the effect of finite size of a network on its percolation threshold. Clerc *et al.* (1980) studied this effect by varying the thickness ℓ of a system made of layers of particles with a diameter R_p. Figure 12.4 shows their results for various thicknesses, from which we obtain $\nu_p{}^3 \simeq 0.85$, in very good agreement with $\nu_p \simeq 0.88$ for three-dimensional percolation (Chapter 2). The finite thickness of the sample also affects the conductivity of the system. Using (12.2) it is not difficult to show that

$$\frac{g_{e1}}{g_{e2}} \sim \left(\frac{\ell_2}{\ell_1}\right)^{(\mu_3 - \mu_2)/\nu_p{}^3}, \tag{12.6}$$

where $g_{e1}(g_{e2})$ is the conductivity of a sample of thickness $\ell_1(\ell_2)$, and $\mu_2(\mu_3)$ is the critical exponent of conductivity for two-dimensional (three-dimensional) percolation. The experimental data of Clerc *et al.* (1980) for the conductivity of their system was also in agreement with (12.6).

What is the effect of a particle size distribution on the transport properties? If, instead of using particles of the same size, we use particles of various diameters, the resulting packing will be disordered compared with a packing of monosize particles. Harris (1974) developed a criterion according to which we can determine whether introduction of relatively weak disorder into a system can change its critical behavior. According to his criterion, if in a d-dimensional percolation system the quantity $\alpha_p = 2 - \nu_p d$ is positive, then introduction of disorder can change the critical behavior of the system. Otherwise, the critical exponents will remain unchanged. Since $\alpha_p(d = 2) = -2/3$ and $\alpha_p(d = 3) \simeq -0.64$, the conclusion is that introduction of a particle size distribution does not change the critical properties of a packing of particles. This has been confirmed in several experiments.

12.2 DC conductivity and percolation properties of metal–insulator films

Thin metal films are two-phase mixtures of metal and nonmetal components. They have interesting structural and transport properties. Their electrical conductivity varies continuously as the volume fraction of the metallic phase is changed. If the volume fraction of the metallic phase is large enough, the metal phase forms a sample-spanning cluster in which the nonmetallic phase is dispersed in it in the form of isolated islands. In this regime, the electrical conductivity of the system is large (or it is *metallic*) and its temperature coefficient of resistance is positive. If the volume fraction of the metallic phase is close to its percolation threshold, we have a complex system of isolated metallic islands, a tortuous and sample-spanning cluster of the metallic component, and islands of the nonmetallic component. If the nonmetallic phase forms a sample-spanning cluster, the system is in the *dielectric* regime, and its electrical conductivity is very small. Transmission electron microscopy (TEM) shows that thin metal films have a discontinuous structure in which the metallic phase constitutes only a fraction of the total volume of the system. Such films have a nearly two-dimensional structure and are formed in the early stages of film growth by a variety of techniques such as evaporation or sputtering. The metallic phase first forms isolated islands that at later stages of the process join together and form a continuous film. Generally speaking, a metal film is considered as thin if its thickness l is less than the percolation correlation length ξ_p of the three-dimensional film.

Thick films of metallic and nonmetallic components, i.e., those with a thickness greater than ξ_p, also have interesting properties and have been studied for a long time. They can also be produced by cosputtering or coevaporation of two components that are insoluble in each other; the nonmetallic component and the metallic phase. They have been studied for their novel superconducting and magnetic properties. However, the discovery by Abeles *et al.* (1975) that the electrical conductivity of such films near p_c follows (12.2) made them the subject of many studies. Thick films are mainly composed of a metal with a bulk lattice structure. The nonmetallic phase is amorphous and often in the form of isolated islands. They are often called *granular* metal films because inspection of their structure shows that the metal grains are often surrounded by very thin amorphous layers of the nonmetallic component, so that metallic grains remain separated. Granular composites always have high percolation thresholds, which can be close to the random close packing fractions, if the grain size is constant and the insulator is relatively thick. Examples of such composites are $Ni–SiO_2$, Al–Si, and Al–Ge. Unlike thin metal films, we cannot easily find clusters of metallic grains in granular metals, therefore it is more difficult to interpret their structure in terms of percolation properties. There are also strong correlations in the structure of thick metal films, so that the random percolation model discussed so far may not be totally suitable. Deutscher *et al.* (1978) proposed that thick metal films have a granular structure if their nonmetallic phase is amorphous, but should be considered as random composites if the insulating phase is crystalline. By random we mean a system in which the two phases cannot be identified from their structures, so that they both play a symmetric role. Let us now discuss several properties of metal films and relate them to percolation quantities.

The first application of percolation to the interpretation of various properties of very thin (two-dimensional) metal films appears to have been made by Liang *et al.* (1976). They prepared ultrathin films which were 1.5 mm wide with preevaporated indium electrodes spaced 25 mm apart. The substrate was SiO film deposited on a glass microscope slide. Several kinds of metallic compounds were evaporated and studied. They found that the sudden drop of the resistivity of semimetal bismuth ultrathin films was very steep, signaling a percolation transition. When they plotted the conductivity of the system versus the area fraction ϕ of the conducting phase, they found that near the percolation threshold the conductivity follows (12.2) with $\mu \simeq 1.15$, reasonably close to $\mu \simeq 1.3$ for two-dimensional percolation discussed in Chapter 2. However, they also found that $\phi_c \simeq 0.67$, quite different from $\phi_c = 0.5$, the prediction of Scher and Zallen (1970) for two-dimensional continuum percolation (Chapter 2). The difference can be attributed to the existence of correlations that are usually present in such films, but were absent in the Scher–Zallen theory for random and continuum percolation.

Beautiful studies of the morphological properties of thin metallic films were carried out by Voss *et al.* (1982) and Kapitulnik and Deutscher (1982), the results of which were published in two back-to-back papers in *Physical Review Letters.* Voss *et al.* prepared Au films by electron beam evaporation onto 30 nm thick amorphous Si_3N_4 windows grown on an Si wafer frame. A range of samples were prepared simultaneously which were from 6 to 10 nm thick and varied from electrically insulating to conducting. Transmission electron micrographs were digitized, and with the use of threshold detection and a connectivity-checking algorithm the individual Au clusters were isolated and their statistical properties were analyzed. Figure 12.5 shows one of their samples at a fractional Au coverage $p = 0.71$, just below the percolation threshold of the Au phase, $p_c \simeq 0.74$. Voss *et al.* found that the Au clusters were irregularly shaped and ramified. At large scales most of the film properties were uniform, but at small scales they were not. Since the deposited Au atoms have some initial mobility, the Au clusters were not totally random, as a result of which the percolation threshold of the system, $p_c \simeq 0.74$, was larger than 0.5, the expected value for two-dimensional continuum percolation. But at p_c Voss *et al.* found that the largest Au

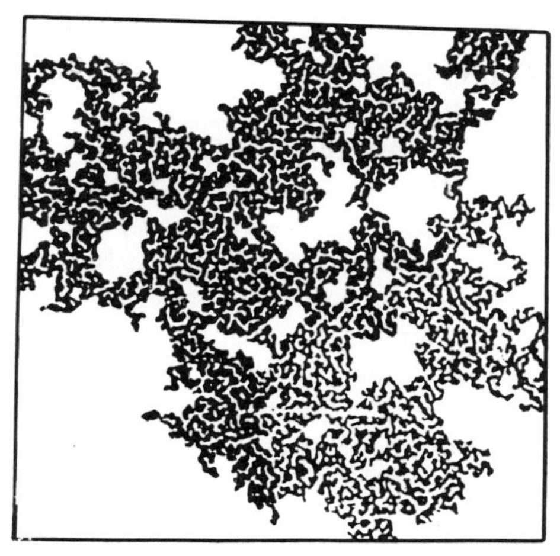

Figure 12.5 *Au clusters at* $p = 0.71$ *(after Voss et al. 1982).*

cluster is a fractal object with a fractal dimension of about 1.9, in excellent agreement with that of two-dimensional percolation (Chapter 2), $D_c = 91/48 \simeq 1.896$.

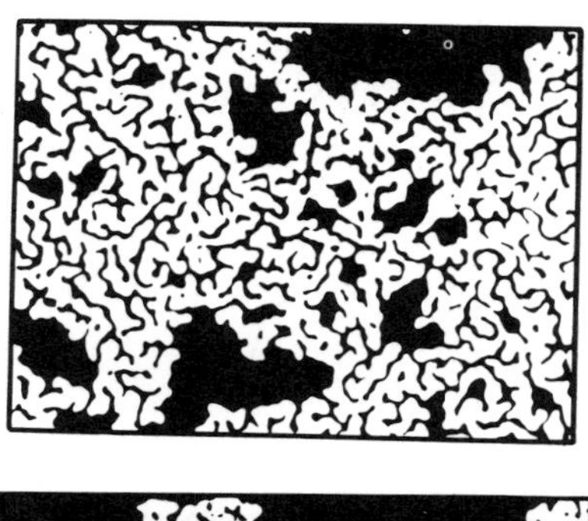

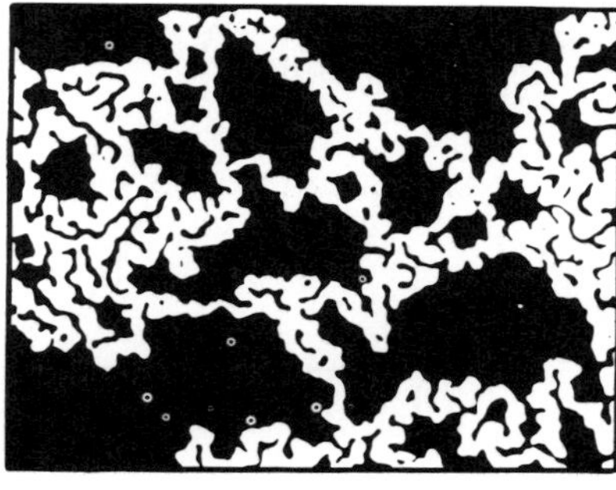

Figure 12.6 *The sample-spanning cluster (top) and its backbone (bottom) of Pb films (after Kapitulnik and Deutscher 1982).*

Kapitulnik and Deutscher (1982) prepared samples by successive deposition of amorphous Ge as the substrate, and of thin Pb films as the metal. This allowed them to obtain deposition with only short-range correlations. The size of the Pb crystallites as the average Pb thickness was about 200 Å. Since the continuity of the metallic cluster is controlled by its thickness, varying the sample thickness is equivalent to generating percolation networks with varying fractions of conducting components. Figure 12.6 shows part of the sample-spanning cluster of Pb and its backbone. Kapitulnik and Deutscher (1982) deposited the thin films on TEM grids and photographed them in the TEM with a very large magnification. The pictures were then analyzed for various percolation properties. Figure 12.7 shows their results for the mass M of the sample-spanning cluster and its backbone. Over about one order of magnitude the sample-spanning cluster and its backbone show fractal behavior, with $D_c \simeq 1.9$ for the cluster and $D_{BB} \simeq 1.65$ for the backbone, in excellent agreement with the results for two-dimensional percolation (Chapter 2). Kapitulnick and Deutscher (1982) also calculated

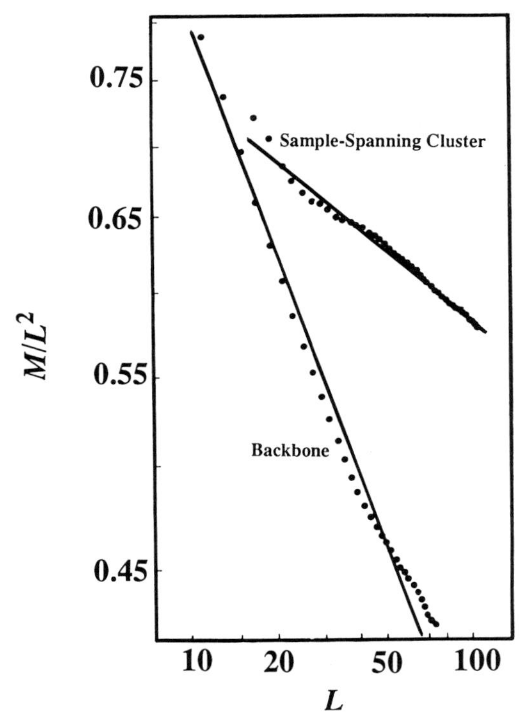

Figure 12.7 *The mass M of the sample-spanning cluster and its backbone of Pb films as a function of the linear size L of the film (after Kapitulnik and Deutscher 1982).*

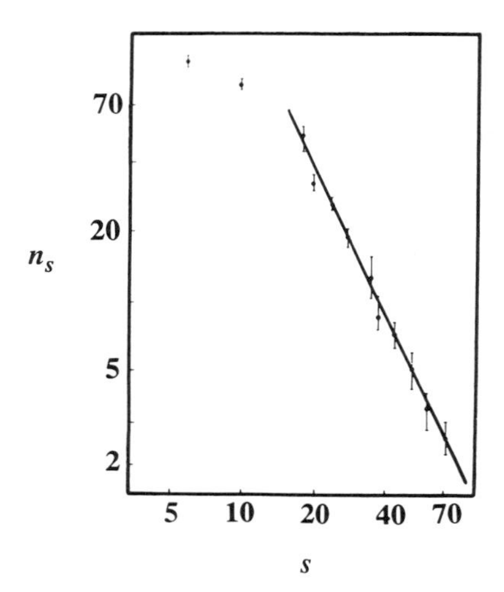

Figure 12.8 *The number of clusters n_s of size s in Pb films (after Kapitulnik and Deutscher 1982).*

n_s, the number of metallic clusters of mass s below p_c. Figure 12.8 shows their results. Equation (2.14) tells us that

$$n_s \sim s^{-\tau_p}, \tag{12.7}$$

and from Fig. 12.8 we obtain $\tau_p \simeq 2.1 \pm 0.2$, completely consistent with $\tau_p = 187/91 \simeq 2.054$ for two-dimensional percolation. Qualitatively similar results were obtained by Laibowitz *et al.* (1982). They prepared thin Al–Al_2O_3 films by thermal evaporation and calculated the size distribution of the metallic regions. They found that this distribution obeys (12.7), although their estimate of τ_p was somewhat lower than the prediction of percolation.

Papandreou and Nédellec (1992) prepared Pd films with a typical thickness of about 100 Å. They deposited them using electron gun evaporation on a quartz substrate for conductivity measurements and on an NaCl substrate for TEM observations. The substrates were coated with a SiO layer of 150 Å, which ensured similar growth conditions, regardless of the substrate below SiO. The samples were then irradiated under normal incidence with a 100keV Xe ion beam, after which their conductivities were measured. They found that as the percolation threshold was approached, the conductivity of the sample vanished according to (12.2) with $\mu \simeq 1.3$, in complete agreement with the prediction of two-dimensional percolation.

Let us now discuss some of the experimental results for thick films. In a seminal paper, Abeles *et al.* (1975) studied the growth of the grains of the metals W or Mo in the insulators Al_2O_3 or SiO_2 using cosputtering. They changed the volume fraction of the metal in the mixture in the range $0.1 < \phi \leqslant 1$ and obtained very finely dispersed grain structures. With

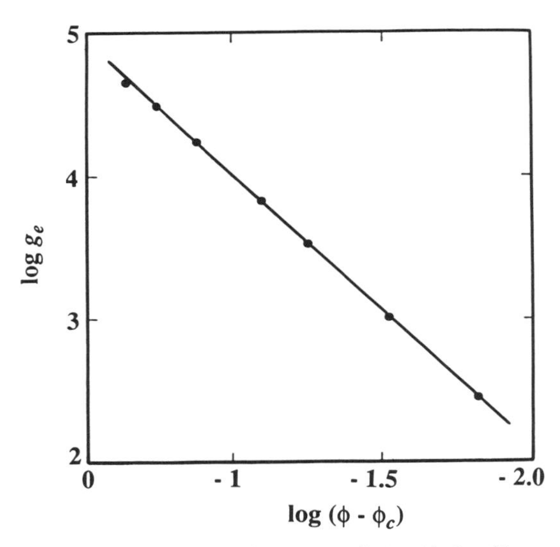

Figure 12.9 *Log–log plot of the conductivity of W–Al_2O_3 films versus the volume fraction* ϕ *of W, where* ϕ_c *is the percolation threshold (after Abeles et al. 1975).*

annealing, we can change the size of the metal grains over a wide range. The advantage of using the W or Mo is that annealing does not cause the metal to precipitate. The metal remains uniformly dispersed within the insulator. The conductivity of the films was determined at various values of the metallic phase volume fraction. Because of correlations, the metal–insulator transition occured at $\phi_c \simeq 0.47$, considerably higher than the Scher–Zallen prediction for three-dimensional continuum percolation, $\phi_c \simeq 0.15$–0.17. Figure 12.9 shows the measured conductivity of the samples prepared by Abeles *et al.* as a function of the volume fraction of the metallic phase. The straight line has a slope $\mu \simeq 1.9 \pm 0.2$, in very good agreement with $\mu \simeq 2.0$ for the critical exponent of three-dimensional percolation conductivity (Chapter 2).

Using percolation networks of resistors, Cohen *et al.* (1978) showed that we may simulate systems that have high percolation thresholds such as granular composites. Their model starts with an initial network already containing a certain fraction p_0 of metallic bonds arranged in a configuration of noncontacting metallic regions. It replaces at random the nonconducting bonds with metallic ones and estimates the conductivity and percolation threshold of the system. In this manner, correlations are introduced into the model, and there is a systematic bias towards the formation of isolated metallic regions, as observed in the films studied by Abeles *et al.* The percolation threshold of the system depends on p_0, and Cohen *et al.* showed that the conductivity data of Abeles *et al.* can be quantitatively predicted by their model. Webman *et al.* (1975) used percolation networks of resistors to predict the electrical conductivity of several types of disordered materials such as metal–amonia solutions, and alkali–

tungsten–bronzes which undergo metal–insulator transitions, similar to those found in the study of Abeles *et al.* (1975). They showed that such percolation networks provide quantitative predictions of the electrical conductivity in all the cases they studied.

Results similar to those of Abeles *et al.* were also reported by Kapitulnik and Deutscher (1983). They prepared Al–Ge films by coevaporating them onto glass substrates from two electron beam guns through a mask with slits, and measured their electrical conductivity. The conductivity of thick samples near the metal–insulator transition obeyed (12.2) with $\mu \simeq 2.1 \pm 0.5$, completely consistent with the prediction of three-dimensional percolation. For thin samples the corresponding exponent was found to be $\mu \simeq 0.9 \pm 0.25$, smaller than for two-dimensional percolation, $\mu \simeq 1.3$, but close to it. The crossover between two- and three-dimensional films were also studied by Kapitulnik and Deutscher (1983). They found that the percolation threshold of the system depended on the thickness of the sample and obeyed (12.4), as expected. More recent experimental data for Al–Ge films are given by Kapitulnik *et al.* (1990), and for In–Ge thin films are given by Tessler and Deutscher (1989).

Optical properties of percolating metal films have also been measured and explained. We do not discuss them here, but refer the interested reader to Yagil and Deutscher (1988), Gadenne and Gadenne (1989), Gadenne *et al.* (1989), and Robin and Souillard (1989) for details.

12.3 AC conductivity and dielectric properties of composite media

In this section we discuss percolation models of AC conductivity and dielectric properties of composite materials. Equation (5.10) shows that EMA can be used for predicting the AC conductivity of disordered systems that are far from their percolation threshold. In this section we focus on the scaling behavior of these properties near p_c. Consider first a regular or random network in which each bond has a conductance equal to either a with probability p, or to b with probability $q = 1 - p$. Using dimensional analysis, we can show the effective conductivity g_e of the network is a homogeneous function and takes the form

$$g_e(p, a, b) = aF(p, h), \tag{12.8}$$

where $h = b/a$ can assume any complex or (positive) real value. By definition the effective conductivity g_e is invariant under the interchange of a and b, and therefore we must have

$$g_e(p, a, b) = g_e(q, b, a) \tag{12.9}$$

$$F(p, h) = hF(q, 1/h). \tag{12.10}$$

Chapter 2 gives two limiting cases of the system; the case in which $b = 0$ and a is finite (conductor–insulator mixtures), and the case in which $a = \infty$ and b is finite (superconductor–conductor mixtures). Both cases correspond to $h = 0$, and therefore the point $h = 0$ at $p = p_c$ is particularly important. Let us therefore focus our attention on this singular point. In the critical region near this point, where both $|p - p_c|$ and h are small, the effective conductivity g_e follows the scaling law

$$g_e \sim a|p - p_c|^{\mu}\Phi_{\pm}(h|p - p_c|^{-\mu - s}), \tag{12.11}$$

where the critical exponents μ and s are defined in Chapter 2, and Φ_+ and Φ_- are two homogeneous functions corresponding, respectively, to the regions above and below p_c. Equation (12.11) was first proposed by Efros and Shklovskii (1976) and Straley (1976). Similar to the exponents μ and s, these scaling functions are universal and do not depend on the network type once h and $p - p_c$ are fixed.

For any fixed and nonzero value of h, the effective conductivity g_e has a smooth dependence on $p - p_c$. This becomes clearer if we rewrite (12.11) in the following form

$$g_e \sim ah^{\mu/(\mu + s)}\psi[|p - p_c|h^{-1/(\mu + s)}], \tag{12.12}$$

where $\psi(x) = x^t\Phi_+(x^{-\mu - s}) = (-x)^t\Phi_-[(-x)^{-\mu - s}]$. The scaling function $\psi(x)$ possesses a Taylor expansion around $x = 0$, $\psi(x) = \psi(0) + \psi_1 x + \psi_2 x^2 + \ldots$, which means that at $p = p_c$ and for $|h| << 1$ we must have

$$g_e \sim \psi(0)(a^s b^{\mu})^{1/(\mu + s)} \equiv \psi(0)ah^u, \tag{12.13}$$

where

$$u = \frac{\mu}{\mu + s}. \tag{12.14}$$

Note that u is the analog of Δ defined in (11.4) for gel polymers. Equation (12.13) implies that $\Phi_-(x) \sim \Phi_+(x) \sim \psi(0)x^u$, which shows clearly the homogeneous nature of these functions. Many other properties of these scaling functions are discussed by Clerc *et al.* (1990), to whom the interested reader is referred.

These results can now be used for modeling the AC conductivity and dielectric properties of a disordered composite near p_c. Many authors have studied this problem, and among the earliest ones we should mention Efros and Shklovskii (1976), and Bergman and Imry (1977). A comprehensive review of this subject is provided by Clerc *et al.* (1990). Here we only summarize the main theoretical results, and discuss their experimental verification. A percolation model can be developed for the AC conductivity and dielectric properties if we view a and b as complex conductances.

Consider a percolation network in which a fraction p of the bonds are purely resistive, while a fraction $q = 1 - p$ of the bonds behave as perfect capacitors. Thus we set

$$a = \frac{1}{R}, \qquad b = iC\omega, \tag{12.15}$$

where R is the resistance of the bond, C is the capacitance, ω is the frequency, and $i = \sqrt{-1}$. The conductance ratio h is then given by

$$h = \frac{i\omega}{\omega_0}, \tag{12.16}$$

where $\omega_0 = 1/(RC)$. In the static limit ($\omega = 0$) the capacitors become insulators, and the model reduces to the usual percolating conductor–insulator mixture already discussed in Chapter 2. This model is usually referred to as the R–C model. The key result is that if p is close to p_c, the R–C model possesses scaling properties, and the effective conductivity $g_e(p, \omega)$ of the system obeys the following scaling law

$$g_e(p, \omega) \sim \frac{1}{R}\,|p - p_c|^{\mu}\,\Phi_{\pm}\left(\frac{i\omega}{\omega_0}\,|p - p_c|^{-\mu - s}\right), \tag{12.17}$$

which follows directly from (12.11). An immediate consequence of (12.17) is the existence of a time scale t_s that diverges as p is approached from *both sides*

$$t_s \sim \omega_0^{-1}\,|p - p_c|^{-(\mu + s)}. \tag{12.18}$$

The significance of t_s is discussed below.

We now define the frequency-dependent complex dielectric constant $\varepsilon(p, \omega)$ of the system by the following equation

$$\varepsilon(p, \omega) = \frac{g_e(p, \omega)}{i\omega}, \tag{12.19}$$

which is generalization of the usual static dielectric constant ε_0. For an insulating dielectric medium $g_e \simeq i\omega\varepsilon_0$ as $\omega \to 0$. By using some general analytic properties of the effective complex dielectric constant of a random mixture, Bergman and Imry (1977) derived the following scaling relations, which can be obtained from (12.17) and (12.19)

$$g_e(p_c, \omega) \sim \omega^{x}, \tag{12.20}$$

$$\varepsilon(p_c, \omega) \sim \omega^{-y}, \tag{12.21}$$

where the exponents x and y satisfy the following relation

$$x + y = 1. \tag{12.22}$$

Equation (12.22) is a direct consequence of the fact that a complex conductivity is an analytic function of $i\omega$. Bergman and Imry (1977) argued that the main contribution to the AC properties is due to polarization effects between various percolation clusters in the disordered medium. Based on this argument, they proposed that

$$x = \frac{\mu}{\mu + s}, \tag{12.23}$$

$$y = \frac{s}{\mu + s}, \tag{12.24}$$

so that in two dimensions where $\mu = s$, we have $x = y = 1/2$. Gefen *et al.* (1983), who studied fractal diffusion on percolation clusters (see Chapter 9), argued that the fractal nature of diffusion and of percolation clusters at length scales up to ξ_p dominate the contributions to the AC properties. They proposed instead that

$$x = \frac{\mu}{\nu_p(2 + \theta)}, \tag{12.25}$$

$$y = \frac{2\nu_p - \beta_p}{\nu_p(2 + \theta)}, \tag{12.26}$$

which also obey (12.22). We shall discuss the experimental verification of these equations later.

An important consequence of (12.17) is the scaling behavior of ε_0 in the critical region near p_c. Using the Taylor expansion of $\Phi_\pm$ discussed above, we can show

$$\varepsilon_0 \sim A_\pm C\,|p - p_c|^{-s}, \tag{12.27}$$

that is, the static dielectric constant *diverges* as p_c is approached from *both sides*; the critical exponent that characterizes this divergence is s, the critical exponent of a percolation network of superconductor–conductor bonds discussed in Chapter 2. This spectacular result was first derived by Efros and Shklovskii (1976). In (12.27), A_+ and A_- represent the amplitudes of ε_0 above and below p_c, respectively. Although (12.27) is supposed to be valid in the limit $\omega = 0$, its validity extends to higher frequencies as long as $\omega << 1/t_s$, where t_s is given by (12.18). An important property of the amplitudes $A_\pm$ is that their ratio A_+/A_- is a universal quantity, independent of the microscopic details of the system. If we now write $g_e(p, \omega) = g' + ig'' = i\omega\varepsilon(p, \omega) = i\omega(\varepsilon' - i\varepsilon'')$, then a loss angle can be defined by

$$\tan\delta = \frac{g'}{g''} = \frac{\varepsilon''}{\varepsilon'}, \tag{12.28}$$

and it is clear that $0 \leqslant \delta \leqslant \pi$. This loss angle is defined in a manner completely analogous to that of gelling solutions discussed in Chapter 11.

Consider now the effective conductivity of the R–C model at p_c. Equation (12.13) tells us that

$$g_e(p, \omega) \sim \frac{\psi(0)}{R}\left(\frac{i\omega}{\omega_0}\right)^u, \tag{12.29}$$

so that at p_c the loss angle δ_c is *universal* and is given by

$$\delta_c = \frac{\pi}{2}(1-u) = \frac{\pi}{2}\frac{s}{\mu+s}, \tag{12.30}$$

which is completely analogous to (11.4) for gelling solutions. Although (12.30) is supposed to be valid exactly at $p = p_c$, it is important to remember that the universal loss angle δ_c can also be observed in a broad frequency range if $|p - p_c|$ is small enough, which implies that $1/t_s << \omega << \omega_0$. Since for *any* two-dimensional system we have $s = \mu$, we must have $\delta_c = \pi/4$, another remarkable result.

Let us now discuss the experimental verification of these percolation predictions for the AC conductivity and dielectric constant. One of the earliest experimental studies of dielectric properties was reported by Castner *et al.* (1975). They measured the static dielectric constant of *n*-type silicon. About 1200 Å of Au was evaporated on thin disk samples which consisted of two imperfect Schottky barriers with thin (about 5–10 Å thick) oxide barriers bounding from 0.2 to 1.0 mm of bulk semiconductor. They varied the concentration c_d of the donor, and showed that ε_0 diverges as c_d approaches a critical concentration from the insulating side. Although percolation was not mentioned in this work, the divergence of ε_0 is a clear indication of the percolation transition indicated by (12.27). To explain these data, Dubrov *et al.* (1976) developed a model in which each bond of a percolation network represented a 300 ohm resistor and a $0.5\,\mu$F capacitor in parallel. Starting with a square network with only capacitors, they added resistors to the network at randomly selected bonds and made measurements of the conductivity of the system at very low frequencies. As the fraction of the resistors approached the percolation threshold of the network, the dielectric constant of the network appeared to diverge. Dubrov *et al.* did write down a scaling law for this divergence that was similar to (12.27), although they did not attempt to estimate the associated critical exponent, since the network they used was too small.

A definitive experimental study of dielectric constant of composite materials near p_c was undertaken by Grannan *et al.* (1981). The composite

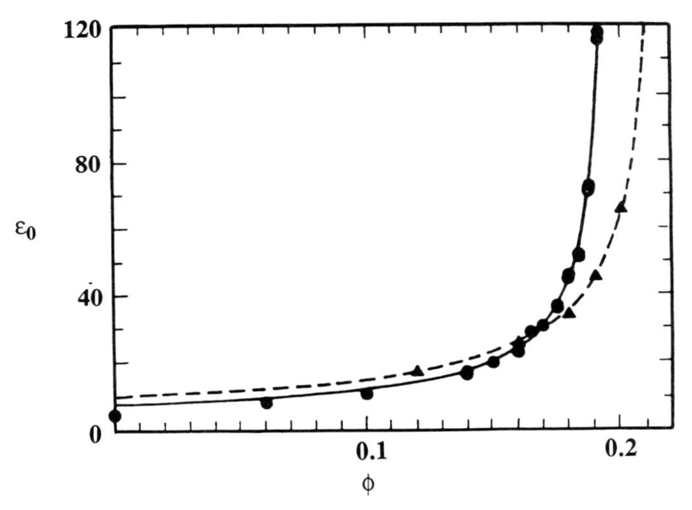

Figure 12.10 *The dielectric constant ε_0 of two series of Ag–KCl composites versus the volume fraction ϕ of Ag. The solid line represents the best fit of the data to (12.27) (after Grannan et al. 1981).*

consisted of small spherical Ag particles randomly distributed in a nonconducting KCl host. The metal particles were prepared by evaporating Ag in the presence of argon gas and a small amount of oxygen. The particles were polydisperse, their sizes varied between 60Å and 600Å, and the overall size distribution was log-normal. The composite was prepared by mixing a given amount of Ag particles and KCl powder and by compressing the mixture into a solid pellet under high pressure. The dielectric constant of the sample was measured by a capacitance bridge operated at 1 kHz. Figure 12.10 shows their results as a function of the volume fraction of Ag. The dielectric constant appears to diverge at $p_c \simeq 0.2$, somewhat larger than $\phi_c \simeq 0.15$–0.17 for three-dimensional percolating continua predicted by Scher and Zallen (1970). Figure 12.11 shows a logarithmic plot of the same data, all of which appear to lie on a straight line indicating that ε_0 diverges as p_c is approached with an exponent $s \simeq 0.73 \pm 0.01$, in perfect agreement with the percolation prediction $s \simeq 0.735$ (see Chapter 2). Similar results were obtained by Nicklasson and Grangvist (1984).

Laibowitz and Gefen (1984) prepared a series of samples of Au films on Si_3N_4 with varying thicknesses selected to span the entire metal–insulator transition. Insulating samples below p_c were easily achievable, indicating that the contribution of tunneling and hopping to the conductivity can be ignored in the more metallic samples. The AC conductivity and capacitance (proportional to the dielectric constant) of the samples were then measured. When the data were fitted to (12.20) and (12.21), they obtained $x \simeq 0.95 \pm 0.05$ and $y \simeq 0.13 \pm 0.05$. These values are in rough agreement

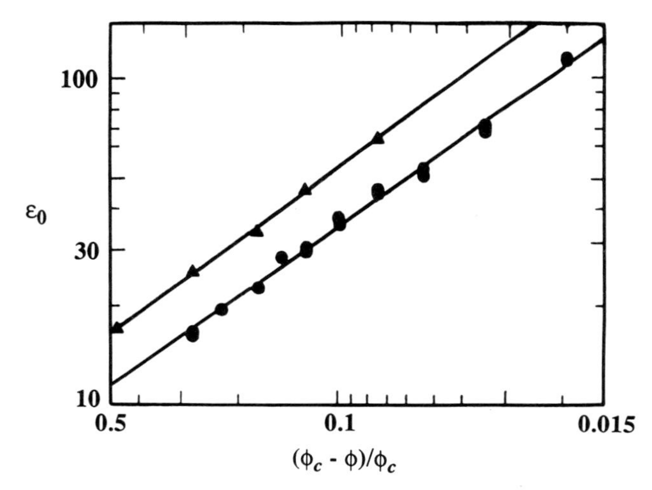

Figure 12.11 *Logarithmic plot of the data shown in Fig. 12.10 (after Grannan et al. 1981).*

with (12.22). But (12.23) and (12.24) predict that for two-dimensional systems (where $\mu = s$) $x = y = 1/2$, while (12.25) and (12.26) predict that $x \simeq 0.34$ and $y \simeq 0.66$, neither of which agree with the experimental data.

Song *et al.* (1986) measured the AC electrical properties of a powder mixture of amorphous carbon and Teflon in the frequency range 10Hz to 13 MHz. Because of its stability, Teflon powder was used as the insulating component. Moreover, the low conductivity of amorphous carbon powder made it possible to easily observe the change of the conductivity as a function of p. The samples were prepared by mixing the carbon and Teflon powder to the desired volume fraction and were then compressed. The electrical conductivity was then measured near p_c and was found to obey (12.2) with $\mu \simeq 1.85 \pm 0.25$, in very good agreement with three-dimensional percolation conductivity. The dielectric constant, in the static limit, was found to diverge according to (12.27) with $s \simeq 0.68 \pm 0.05$, in good agreement with the theoretical expectation. The AC conductivity and the dielectric constant were also measured, from which it was estimated that $x \simeq 0.86 \pm 0.06$ and $y \simeq 0.12 \pm 0.04$ They do not agree with (12.23) and (12.24) which predict that $x \simeq 0.6$ and $y \simeq 0.4$, nor do they agree with the predictions of (12.25) and (12.26), $x \simeq 0.73$ and $y \simeq 0.27$.

The resolution of this apparent disagreement between theory and experimental data was provided by Hundley and Zettl (1988). They measured the AC conductivity and dielectric constant of thin Au films, similar to those of Laibowitz and Gefen (1984), but extended the frequency range to between 100Hz and 1 GHz. Their measurements, shown in Fig. 12.12, indicate that in the intermediate frequency regime, corresponding to that of Laibowitz and Gefen and Song *et al.*, $x \simeq 1.0$ and $y \simeq 0$, in agreement with the data of Laibowitz and Gefen (1984). At higher frequencies $x \simeq 0.32$, in

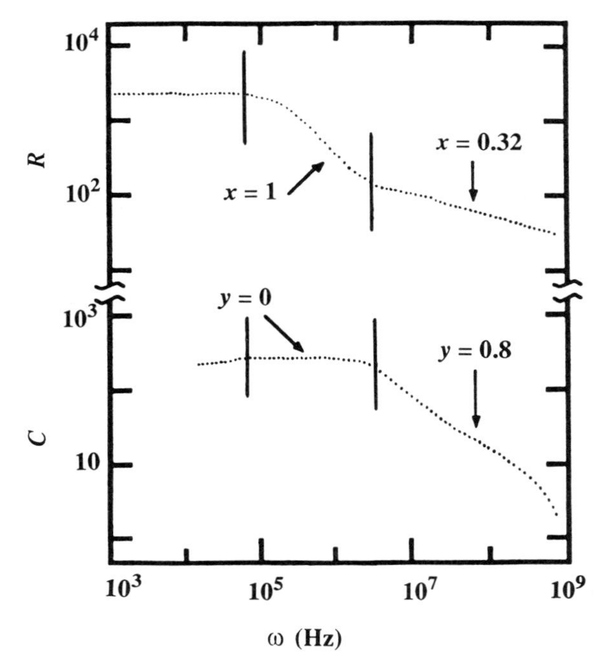

Figure 12.12 *Frequency dependence of the resistance R and capacitance C of an Au fractal film at T = 300 K. The critical exponents x and y are defined by (12.20) and (12.21) (after Hundley and Zettl 1988).*

excellent agreement with the prediction of (12.25), and $y \simeq 0.8$, in rough agreement with the prediction of (12.26). Thus, the fractal nature of percolation clusters seems to play an important role in the AC conductivity and dielectric properties of disordered materials at high frequencies.

Laugier *et al.* (1986) measured AC conductivity of random mixtures of glass microbeads; a varying fraction of them had their surface coated with silver but this did not appreciably change the density of the powder. The average diameter of the beads was about 30 μm and the frequencies used were as large as 50 MHz. The loss angle δ was also measured. Equation (12.30) predicts that at p_c we must have $\tan \delta_c \simeq 0.45$, while the measured value was $\tan \delta_c \simeq 0.5$, in good agreement with the prediction.

Microemulsions are another class of disordered systems whose AC conductivity and dielectric properties have been measured. They are thermodynamically stable, isotropic, and transparent dispersions of two immiscible fluids, such as water and oil, with one or more surfactants that are surface active. A water in oil (W/O) microemulsion usually consists of small spherical water droplets surrounded by a monomolecular layer of surfactant and dispersed in a continuous oil phase. The W/O microemulsions usually have a small macroscopic conductivity because the water droplets are

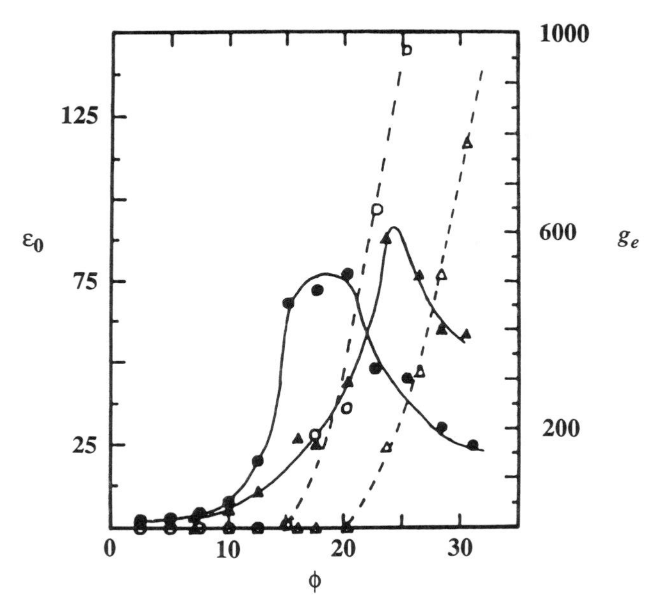

Figure 12.13 *Dielectric constant* ε_0 *and DC conductivity* g_e *of two different micromulsions as a function of the volume fraction* ϕ *of water. The data are for* $W_0 = [H_2O]/[AOT] = 25$ *and* $T = 318K$ *(circles), and for* $W_0 = 35$ *and* $T = 313K$ *(after van Dijk 1985).*

separated by the surfactant layers and the oil phase. Ionic surfactants can donate an ion to the water phase and increase its conductivity. If the volume fraction ϕ_w of the water phase exceeds a critical value ϕ_{wc}, the conductivity increases sharply, usually by several orders of magnitude. This large conductivity rise is due to the fact that charge carriers are able to move along connected paths in the microemulsion. Thus, the conductivity transition in microemulsions is a percolation phenomenon, which can also be induced by increasing the temperature while holding ϕ_w constant.

van Dijk (1985) was probably the first to measure the dielectric constant of a microemulsion at ϕ_{wc}. His system was a microemulsion of AOT, sodium di-2-ethylhexylsulfosuccinate (an anionic surfactant with an SO_3^- head group and two hydrocarbon tails), water and isooctane. The volume fraction of water was changed by varying the amount of oil and keeping the molar ratio water/AOT constant. Figure 12.13 shows the data which indicate a sharp peak for the dielectric constant and a dramatic increase for the electrical conductivity, both at ϕ_{wc}, in agreement with the predictions of percolation. Moreover, (12.30) tells us that at ϕ_{wc} the loss angle δ_c is independent of frequency, and Fig. 12.14 indicates that over more than one order of magnitude this is indeed the case. We also obtain $u \simeq 0.62 \pm 0.02$, reasonably close to the percolation prediction $u = \mu/(\mu + s) \simeq 0.73$. More

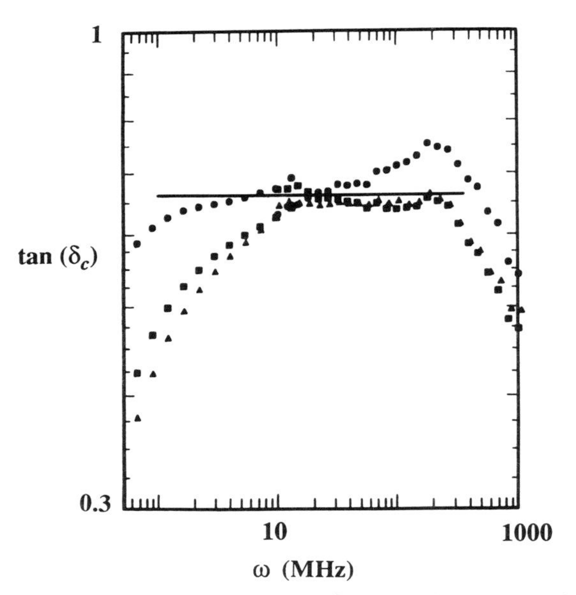

Figure 12.14 *Logarithmic plot of the loss angle* δ_c *versus frequency* ω *for three different microemulsions. The horizontal line indicates* $\tan(\delta_c) = 0.67$ *(after van Dijk 1985).*

extensive measurements for the same microemulsions were reported by van Dijk *et al.* (1986).

Moha-Ouchane *et al.* (1987), Clarkson and Smedley (1988), and Peyrelasse *et al.* (1988) have all measured the AC conductivity and dielectric constant of the same microemulsion system as that of van Dijk *et al.* These authors found that the static dielectric constant diverges at ϕ_{wc} with an exponent $s' \simeq 1.6$, significantly larger than $s \simeq 0.73$, predicted by (12.27). Grest *et al.* (1986) argued that in microemulsions we have to take into account the effect of cluster diffusion that rearranges the system and dynamically changes the structure of the system. They proposed a dynamic percolation model in which the percolation clusters diffuse randomly in the network, very similar to the model of Arbabi and Sahimi discussed in Chapter 11 for the Rouse regime of gelation. Based on this argument, Grest *et al.* (1986) argued that the exponent, s, characterizing the divergence of the static dielectric constant, should be replaced by $s' = 2\nu_p - \beta_p$, identical to (11.36) for the viscosity of gelling solutions in the Rouse limit. Equation (12.26) also implies that s should be replaced with s'. But even $s' \simeq 1.35$ seems to be in disagreement with the measurements of Moha-Ouchane *et al.*, Clarkson and Smedley, and Peyrelasse *et al.*, and the disagreement between the measurement and the predictions remains unexplained. See Clarkson (1988) for an extensive discussion of various properties of microemulsion systems and a comparison between experimental data and the predictions of percolation.

12.4 Hall conductivity of composite media

A popular tool for investigating electrical transport in composite mixtures of good and poor conductors is to look at the properties of the system in the presence of a magnetic field H. Then, the conductivity of the system is described by a tensor $\mathbf{g}$ which has nonzero off-diagonal terms, even if the system is isotropic. Some of the off-diagonal terms are symmetric and even in the magnetic field H, while others are antisymmetric and odd in H. Suppose that an electric field $\mathbf{E}$ has also been imposed on the system. Then, the system has an electrical (ohmic) conductivity g_e, and a *Hall conductivity* $\mathbf{g}_h$. For an isotropic medium and small H, $\mathbf{g}_h$ is proportional to H and is defined by the Kirchhoff's law relating the current density $\mathbf{J}$ and the electric field $\mathbf{E}$

$$\mathbf{J} = g_e\mathbf{E} + \mathbf{E} \times \mathbf{g}_h. \tag{12.31}$$

If (12.31) is inverted for small H, the Hall coefficient $R_h = g_h/(Hg_e^2)$ appears along the ohmic resistivity $\rho = 1/g_e$: $\mathbf{E} = \rho\mathbf{J} + R_h(\mathbf{H} \times \mathbf{J})$. Thus, one main goal has been calculating the Hall conductivity and coefficient in disordered composites, and investigating the scaling behavior of g_h and R_h near p_c. Juretschke *et al.* (1956) pioneered such calculations by treating various types of disorder in three-dimensional systems. They also obtained the exact general solution for the Hall effect at low H in isotropic two-dimensional composites. Cohen and Jortner (1973) developed an EMA for the problem, an extension of the EMA discussed in Chapter 5. Unlike the case of g_e, the presence of the magnetic field makes the development of a network model for calculating g_h and R_h a quite complex task. For example, the general circuit element of a bond of the network cannot be a simple resistor, but has to be a conductance *matrix*. Straley (1980b), Bergman *et al.* (1983), and Duering and Bergman (1989) have proposed various network models for calculating g_h and R_h.

The study of the scaling behavior of g_h and R_h near p_c was initiated by Levinshtein *et al.* (1976), Shklovskii (1977), and Straley (1980a). A complete scaling theory was developed by Bergman and Stroud (1985). Consider a two-component network in which each bond has good ohmic and Hall conductivities σ_1 and σ_{h1} with probability p, and poor ohmic and Hall conductivities σ_2 and σ_{h2} with probability $1 - p$. Bergman and Stroud (1985) proposed that for $\sigma_{h2}/\sigma_{h1} \ll 1$ and for p close to p_c.

$$\frac{g_h - \sigma_{h2}}{\sigma_{h1} - \sigma_{h2}} = |p - p_c|^{\tau} F\left[\frac{\sigma_{h2}/\sigma_{h1}}{|p - p_c|^{\mu + s}}\right], \tag{12.32}$$

where τ is a new critical exponent characterizing the scaling behavior of g_h near p_c when $p > p_c$, $\sigma_2 = 0$, and $\sigma_{h2} = 0$. Equation (12.32) has been written in complete analogy with (12.11). The scaling function $F(x)$ has the properties that, $F(x) \sim$ *constant* if $x << 1$ and $p > p_c$ (Regime I), $F(x) \sim x^2$ for $x << 1$ and $p < p_c$ (Regime II), and $F(x) \sim x^{\tau/(\mu + s)}$ for $x >> 1$ and $p \simeq p_c$ (Regime III).

A better way of understanding (12.32) is in terms of the Hall coefficient R_h. To each bond we assign a Hall coefficient $R_{h1} = \sigma_{h1}/\sigma_1^2$ with probability p, or $R_{h2} = \sigma_{h2}/\sigma_2^2$ with probability $1 - p$. Then, according to the scaling theory of Bergman and Stroud (1985) we have

$$R_h \sim a_1 R_{h1} |p - p_c|^{-g} + b_1 R_{h2}(\sigma_2/\sigma_1)^2 |p - p_c|^{-2\mu}, \qquad (12.33)$$

in Regime I,

$$R_h \sim a_2 R_{h1} |p - p_c|^{-g} + b_2 R_{h2} |p - p_c|^{2s}, \qquad (12.34)$$

in Regime II, and

$$R_h \sim a_3 R_{h1}(\sigma_2/\sigma_1)^{-g/(\mu + s)} + b_3 R_{h2}(\sigma_2/\sigma_1)^{2s/(\mu + s)}, \qquad (12.35)$$

in Regime III, where a_i and b_i are constant, and g is a critical exponent that characterizes the divergence of R_h as p_c is approached from below, and is related to μ and τ by

$$g = 2\mu - \tau. \qquad (12.36)$$

It is not yet clear whether τ (or g) is an independent exponent, or is related to the other percolation exponents defined so far. Straley (1980a) proposed that

$$g = 2[\mu - (d - 1)\nu_p]. \qquad (12.37)$$

This equation predicts that $g(d = 3) \simeq 0.48$, fully compatible with the numerical estimate $g \simeq 0.49 \pm 0.06$. It also predicts that in the mean field limit $g = 1$, in agreement with the exact calculation of Straley (1980b). Equation (12.37) also predicts that $g(d = 2) \simeq -0.03$, in disagreement with the exact result $g(d = 2) = 0$ obtained by Shklovskii (1977). As pointed out by Bergman and Stroud (1985), the important point to remember is that in Regime I (12.33) the ratio of the second to the first term is of the order $(\sigma_{h2}/\sigma_{h1}) |p - p_c|^{-\tau}$, and therefore, depending on the parameters of the system, either term may dominate, so that the experimental verification of these scaling laws is not straightforward.

One of the first experimental studies of Hall conductivity and coefficient in a composite was carried out by Levinshtein *et al.* (1976). They performed Hall experiments on 20 mm × 60 mm conducting graphite paper sheets with holes randomly punched in them. For three-dimensional experiments a compressed stack of 15 individually punched graphite sheets was used. The results confirmed the divergence of R_h as p_c is approached. Sichel and

Gittleman (1982) measured g_h and R_h in the granular metals Au–SiO_2 and W–Al_2O_3 near their percolation thresholds. Beautiful experimental verification of (12.33)–(12.35) was provided by Palevski *et al.* (1984) who showed that the Hall coefficient does indeed remain finite in a two-dimensional (thin) metal film, and by Uri *et al.* (1987) who studied the problem in three-dimensional Al–Ge metal–insulator films. In Regime I, the poor conductor was Al-doped Ge, which was found to dominate the Hall effect. To make the first term of (12.33) dominant, Uri *et al.* dissolved the metallic aluminum in KOH. This left the doped Ge as the good conductor; the poor conductor role was now played by the vacuum. In this situation, the first term of (12.33) is dominant, therefore the critical exponents g and τ could be measured directly. The results were $\tau \simeq 3.8 \pm 0.2$ and $g \simeq 0.38 \pm 0.05$, which should be compared with accurate simulation results based on percolation networks $\tau \simeq 3.56 \pm 0.06$ and $g \simeq 0.49 \pm 0.06$. Theory and experiment are in agreement with each other. Since the EMA of Cohen and Jortner (1973) can be used for estimating g_h and R_h away from p_c, we now have a fairly complete theory of Hall effects in composite materials.

12.5 Percolation properties of granular superconductors

Percolation theory can also explain some of the observed properties of granular superconductors. These materials are composed of superconducting grains separated by thin insulating grains; they are called Josephson junctions. A granular superconductor can be characterized by two parameters: the size of the grains and the energy barrier between the grains. The grains size distribution is measured by electron microscopy, while the properties of the barriers are deduced from measurements of normal state (nonsuperconducting) resistivity. These two parameters determine the Josephson energy coupling E_j. Garland (1989) has reviewed many granular properties of superconducting materials.

Deutscher *et al.* (1980) proposed a percolation model for the onset of superconductivity. They assumed that the grains are coupled if $E_j > k_b T$, where k_b is the Boltzman constant, and T is the temperature. Because the coupling energy depends on T, more and more grains become coupled as T is lowered. The coupling was obtained randomly with a probability that depended on T. This defines a percolation process, because a sample-spanning cluster of coupled superconducting grains is formed when the coupling probability is equal to the percolation threshold. If the temperature is still lowered, the sample-spanning cluster of the superconducting grains grows in size. The distribution of the grain sizes and the junction resistances give rise to randomness in the coupling energy. In their model, the coupling between the grains can be very strong, in which case electron microscopy

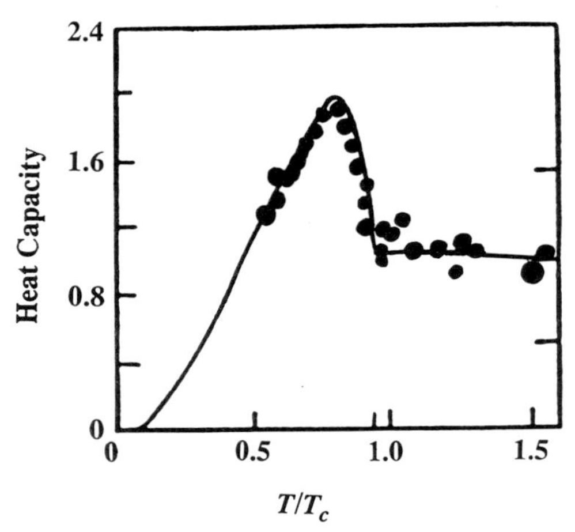

Figure 12.15 *Comparison of percolation prediction of temperature dependence of the heat capacity of Al–Al_2O_3 films (curve) with the experimental data (circles) (after Deutscher et al. 1980).*

shows that the grain size distribution is quite broad. In this case the randomness in the coupling energy is due to the dependence on the grain size of the temperature at which a grain becomes superconducting. Alternatively, if the grain size distribution is narrow, the coupling can be weak. The model of Deutscher *et al.* was developed for weak coupling. They used a dense random packing of hard spheres in which each sphere represented a grain. The sphere sizes were about 30Å, small enough that only the sample-spanning cluster of the superconducting grains was the main contributor to the specific heat of the system, the main property that was calculated. The coupling energy was calculated by assuming that the normal state resistances of the junctions obey a normal distribution whose variance was assumed to be proportional to its mean. The mean of the distribution was treated as the only free parameter of the model. Using this model, the temperature dependence of the specific heat of the system was calculated. Figure 12.15 compares the percolation predictions and the experimental data for granular Al–Al_2O_3 films; the agreement is excellent.

Several other properties of granular superconductors can also be predicted by percolation. Although there is Josephson tunneling through the junctions, it does not generate any potential difference between the grains up to a *critical current* J_c. If the temperature of the system is decreased below the critical temperature T_c of the system, the number of superconducting links N_ℓ (per unit cross-sectional area) increases, and percolation theory can be used for estimating N_ℓ, the critical current J_c, and the critical temperature T_c at which the material becomes superconducting. In the percolation model of Deutscher *et al.* (1980) T_c is close to the temperature at which the specific heat attains its maximum (see Fig. 12.15). Alternatively, T_c can be defined as the temperature at which 50% transition in the resistivity occurs. As another example, consider the scaling behavior of J_c. Near p_c we have the scaling law

$$J_c \sim (p - p_c)^{\upsilon}, \tag{12.38}$$

where υ is a new critical exponent. To relate υ to the other percolation exponents we argue as follows. Near p_c, the granular superconductor can be viewed as a network of superconducting blobs connected by links (see Chapter 2). The links are quasilinear chains of Josephson junctions. The current that flows through parallel links is inversely proportional to N_ℓ, and the distance between the links is roughly proportional to the percolation correlation length ξ_p. For a d-dimensional system, N_ℓ is of the order $1/\xi_p^{d-1}$. Since $J_c \sim N_\ell$, we obtain

$$J_c \sim (p - p_c)^{\upsilon} \sim (p - p_c)^{\nu_p(d-1)}, \tag{12.39}$$

so that υ may be a purely topological exponent. This prediction was confirmed by Monte Carlo simulations in two-dimensional percolation networks (Kirkpatrick 1979, Lobb and Frank 1979), and by careful experiments of Deutscher and Rappaport (1979) who prepared thin (two-dimensional) superconductor–semiconductor films of Pb–Ge, and measured their critical currents. They obtained $\upsilon \simeq 1.3 \pm 0.1$, in very good agreement with the prediction $\upsilon = \nu_p = 4/3$.

How does a magnetic field affect the behavior of a granular superconductor? There are two types of superconductors that behave quite differently in an external magnetic field. Type I superconductors remain superconducting and expel the magnetic field up to a critical field H_c. They then abruptly cross over to the normal state. Type II superconductors allow the magnetic field to penetrate into the system if H is greater than a lower critical field H_{c1} but lose their superconductivity at an *upper critical field* H_{c2}. A characteristic length scale of granular superconductors is their *superconducting coherence length* ξ_s, which can be defined as the diffusion length over a characteristic time t_c for the relaxation of the order parameter at a given T. Therefore, we have $\xi_s \sim \sqrt{Dt_c}$, where D is a diffusion coefficient. Moreover, t_c is given by $t_c \sim (T_c - T)^{-1}$. The homogeneous regime is defined by $\xi_s > \xi_p$, and $H_{c2} \sim (T_c - T)$ near T_c. The upper critical field of a Type II superconductor is proportional to ξ_s^{-2}. In the inhomogeneous regime, $\xi_s << \xi_p$, we have fractal diffusion (Chapter 9), instead of normal diffusion. Since ξ_s is defined as a diffusion length scale, ξ_s^2 should be proportional to $\langle r^2 \rangle$, the mean-square displacement of a random walker given by (9.3) and (9.5). Therefore, we obtain $H_{c2} \sim \xi_s^{-2} \sim t_c^{-2/(2+\theta)}$, which implies that

$$H_{c2} \sim (T_c - T)^{2/(2+\theta)}, \tag{12.40}$$

where $\theta = (\mu - \beta_p)/\nu_p$, defined in Chapter 9. Equation (12.40) was confirmed experimentally by Gerber and Deutscher (1989). They measured the upper critical field of thin semicontinuous lead films on germanium substrates and found that $H_{c2} \sim (T_c - T)^{\gamma}$ with $\gamma = 0.66 \pm 0.06$. Since in two dimensions $\theta \simeq 2.87$, the percolation prediction $\gamma = 2/(2+\theta) \simeq 0.7$ is in very good agreement with the measured value.

A final issue that can be investigated with the aid of percolation is the crossover between percolation and the classical Anderson localization, relevant to the conductivity and the onset of superconductivity in thin metal–insulator films. Deutscher *et al.* (1985) introduced two parameters. One of them is $z = \xi_l/\xi_p$, where ξ_l is the *localization correlation length*, which diverges as $(p - p_c)^{-\nu_l}$ and is different from the percolation correlation length ξ_p. The other is $w = \rho_0 e^2/(L\hbar)$, where ρ_0 is the resistivity of the sample at a minimum length scale L over which it can be defined, e is the electron charge, and $\hbar$ is the Planck's constant. Using the ideas of Abrahams *et al.* (1979) regarding the location of Anderson localization, Deutscher *et al.* (1985) showed that $z = w^{\nu_p/(\mu - \nu_p)}(p - p_c)^{-\nu_l}$ and proposed the following scenario for the crossover. (1) If $z << 1$, then percolation dominates and the resistance ρ of the sample is given by the usual scaling law, $\rho \sim (p - p_c)^{-\mu}$. (2) If $z >> 1$, then localization effects are dominant and $\rho \sim (p - p_c)^{-\nu_l}$. Deutscher *et al.* (1985) showed that this scenario is completely consistent with the experimental data. Thus, the relation between percolation and classical localization was clarified.

12.6 Conclusions

Percolation provides a quantitative description of morphological properties, electrical and Hall conductivities, elastic moduli, and dielectric constants of a wide variety of disordered media and materials, ranging from powders and polymer composites, to metal–insulator films, granular superconductors, and microemulsions. The main prediction of percolation, universal scaling laws such as (12.2) and (12.3), are valid only in the critical region. This region is supposed to be near p_c but is often very broad. So scaling laws can be very useful for quantitative modeling of the effective properties of disordered materials as well as providing insights into the structure of the system and mechanisms for transport.

References

Abeles, B., Pinch, H.L., and Gittleman, J.I., 1975, *Phys. Rev. Lett.* **35**, 247.
Abrahams, E., Anderson, P.W., Licciardello, D.C., and Ramakrishnan, T.V., 1979, *Phys. Rev. Lett.* **42**, 673.
Balberg, I., Binenbaum, N., and Bozowski, S., 1983, *Solid State Commun.* **47**, 989.

Bergman, D.J. and Imry, Y., 1977, *Phys. Rev. Lett.* **39**, 1222.
Bergman, D.J., Kantor, I., Stroud, D., and Webman, I., 1983, *Phys. Rev. Lett.* **50**, 1512.
Bergman, D.J. and Stroud, D., 1985, *Phys. Rev. B* **32**, 6097.
Castner, T.G., Lee, N.K., Cieloszyk, G.S., and Salinger, G.L., 1975, *Phys. Rev. Lett.* **34**, 1627.
Clerc, J.P., Giraud, G., Alexander, S., and Guyon, E., 1980, *Phys. Rev. B* **22**, 2489.
Clerc, J.P., Giraud, G., Laugier, J.M., and Luck, J.M., 1990, *Adv. Phys.* **39**, 191.
Clerc, J.P., Giraud, G., and Roussenq, J., 1975, *C.R. Acad. Sci. Paris B* **281**, 227.
Clarkson, M.T., 1988, *Phys. Rev. A* **37**, 2079.
Clarkson, M.T. and Smedley, S.I., 1988, *Phys. Rev. A* **37**, 2070.
Cody, G.D., Geballe, T.H., and Sheng, P. (eds.), 1990, *Physical Phenomena in Granular Materials*, Pittsburgh, Materials Research Society.
Cohen, M.H. and Jortner, J., 1973, *Phys. Rev. Lett.* **30**, 696.
Cohen, M.H., Jortner, J., and Webman, I., 1978, *Phys. Rev. B* **17**, 4555.
Deprez, N., McLachlan, D.S., and Sigalas, I., 1989, *Physica* **A157**, 181.
Deptuck, D., Harrison, J.P., and Zawadzki, P., 1985, *Phys. Rev. Lett.* **54**, 913.
Deutscher, G., Entin-Wohlman, O., Fishman, S., and Shapira, Y., 1980, *Phys. Rev. B* **21**, 5041.
Deutscher, G., Goldman, A.M., and Micklitz, H., 1985, *Phys. Rev. B* **31**, 1679.
Deutscher, G. and Rappaport, M.L., 1979, *J. Physique Lett.* **40**, L219.
Deutscher, G., Rappaport, M.L., and Ovadyahu, Z., 1978, *Solid State Commun.* **28**, 593.
Domb, C. and Sykes, M.F., 1961, *Phys. Rev.* **122**, 77.
Dubrov, V.E., Levinshtein, M.E., and Shur, M.S., 1976, *Sov. Phys. -JETP* **43**, 1050.
Duering, E. and Bergman, D.J., 1989, *Physica* **A157**, 125.
Efros, A.L. and Shklovskii, B.I., 1976, *Phys. Status Solidi b* **76**, 475.
Fitzpatrick, J.P., Malt, R.B., and Spaepen, F., 1974, *Phys. Lett.* **47A**, 207.
Gadenne, M. and Gadenne, P., 1989, *Physica* **A157**, 344.
Gadenne, P., Yagil, Y., and Deutscher, G., 1989, *Physica* **A157**, 279.
Garland, J.C. and Tanner, D.B. (eds.), 1978, *AIP Conference Proceedings* **40**.
Garland, J.C., 1989, *Physica* **A157**, 111.
Gefen, Y., Aharony, A., and Alexander, S., 1983, *Phys. Rev. Lett.* **50**, 77.
Gerber, A. and Deutscher, G., 1989, *Phys. Rev. Lett.* **63**, 1184.
Grannan, D.M., Garland, J.C., and Tanner, D.B., 1981, *Phys. Rev. Lett.* **46**, 375.
Grest, G.S., Webman, I., Safran, S.A., and Bug, A.L.R., 1986, *Phys. Rev. A* **33**, 2842.
Harris, A.B., 1974, *J. Phys. C* **7**, 1671.
Hsu, W.Y. and Berzins, T., 1985, *J. Polym. Sci. Polym. Phys. Ed.* **23**, 933.
Hundley, M.F. and Zettl, A., 1988, *Phys. Rev. B* **38**, 10290.
Jurutschke, H.J., Landauer, R., and Swanson, J.A., 1956, *J. Appl. Phys.* **27**, 838.
Kapitulnik, A. and Deutscher, G., 1982, *Phys. Rev. Lett.* **49**, 1444.
Kapitulnik, A. and Deutscher, G., 1983, *J. Phys. A* **16**, L255.
Kapitulnik, A., Hsu, J.W.P., and Hahn, M.R., 1990, in *Physical Phenomena in Granular Materials*, edited by G.D. Cody, T.H. Geballe, and P. Sheng, Pittsburgh, Materials Research Society, p. 153.
Kirkpatrick, S., 1979, *AIP Conference Proc.* **58**, 340.
Lafait, J. and Tanner, D.B. (eds.), 1989, *Electrical Transport and Optical Properties of Inhomogeneous Media II, Physica* **A157**, 1–676.
Laibowitz, R.B., Allessandrini, E.I., and Deutscher, G., 1982, *Phys. Rev. B* **25**, 2965.
Laibowitz, R.B. and Gefen, Y., 1984, *Phys. Rev. Lett.* **53**, 380.
Laugier, J.M., Clerc, J.P., and Giraud, G., 1986, *Proceedings of International AMSE Conference*, edited by G. Mesnard, Lyon, AMSE.
Lee, S.-I., Song, Y., Noh, T.W., Chen, X.-D., and Gaines, J.R., 1986, *Phys. Rev. B* **34**, 6719.

Levinshtein, M.E., Shur, M.S., and Efros, A.L., 1976, *Sov. Phys.-JETP* **42**, 1120.
Liang, N.T., Shan, Y., and Wang, S.-Y., 1976, *Phys. Rev. Lett.* **37**, 526.
Lobb, C.J. and Frank, D.J., 1979, *AIP Conference Proc.* **58**, 308.
Malliaris, A. and Turner, D.T., 1971, *J. Appl. Phys.* **42**, 614.
Moha-Ouchane, M., Peyrelasse, J., and Boned, C., 1987, *Phys. Rev. A* **35**, 3027.
Nicklasson, G.A. and Grangvist, C.G., 1984, *J. Appl. Phys.* **55**, 3382.
Ottavi, H., Clerc, J.P., Giraud, G., Roussenq, J., Guyon, E., and Mitescu, C.D., 1978, *J. Phys. C* **11**, 1311.
Palevski, A., Rappaport, M.L., Kapitulnik, A., Fried, A., and Deutscher, G., 1984, *J. Physique Lett.* **45**, L367.
Papandreou, N. and Nédellec, P., 1992, *J. Physique. I France* **2**, 707.
Peyrelasse, J., Moha-Ouchane, M., and Boned, C., 1988, *Phys. Rev. A* **38**, 904.
Robin, T. and Souillard, B., 1989, *Physica* **A157**, 285.
Scher, H. and Zallen, R., 1970, *J. Chem. Phys.* **53**, 3759.
Sichel, E.K. and Gittleman, J.I., 1982, *Solid State Commun.* **42**, 75.
Shklovskii, B.I., 1977, *Sov. Phys.-JETP* **45**, 152.
Shklovskii, B.I., 1978, *Phys. Status Solidi b* **85**, k111.
Smith, L.N. and Lobb, C.J., 1979, *Phys. Rev. B* **20**, 3653.
Song, Y., Noh, T.W., Lee, S.-I., and Gaines, J.R., 1986, *Phys. Rev. B* **33**, 904.
Straley, J.P., 1976, *J. Phys. C* **9**, 783.
Straley, J.P., 1980a, *J. Phys. C* **13**, L773.
Straley, J.P., 1980b, *J. Phys. C* **13**, 4335.
Tessler, L.R. and Deutscher, G., 1989, *Physica* **A157**, 154.
Troadec, J.P., Bideau, D., and Guyon, E., 1981, *J. Phys. C* **14**, 4807.
Uri, D., Palevski, A., and Deutscher, G., 1987, *Phys. Rev. B* **36**, 790.
van Dijk, M.A., 1985, *Phys. Rev. Lett.* **55**, 1003.
van Dijk, M.A., Castelejin, G., Joosten, J.G.H., and Levine, Y.K., 1986, *J. Chem. Phys.* **85**, 626.
Voss, R.F., Laibowitz, R.B., and Allessandrini, E.I., 1982, *Phys. Rev. Lett.* **49**, 1441.
Webman, I., Jortner, J., and Cohen, M.H., 1975, *Phys. Rev. B* **11**, 2885.
Yagil, Y. and Deutscher, G., 1988, *Appl. Phys. Lett.* **52**, 373.

13
Hopping conductivity of semiconductors

13.0 Introduction

Hopping conduction in semiconductors was first associated with the observation that the activation energy of the conductivity in doped Ge exhibits a break at low temperatures T. This observation was first made by Hung and Gliessman (1950) who attributed it to a distinct mechanism of conduction at low values of T. Mott (1956) and Conwell (1956) proposed a model of conduction in which electrons conduct by thermally activated tunneling from a filled site to a vacant one, a process that is usually called *phonon-assisted hopping*. This model was modified and extended by several researchers. The best-known extension is perhaps the model of Miller and Abrahams (1960). They developed a model consisting of two parts, the quantum mechanical theory of the wave functions and of the transition rates W_{ij} from a localized state i to a localized state j, and a statistical mechanical theory of transport that employs such transition rates. They also showed how their model can be reduced to a random resistor network and be used for computing the hopping conductivity of disordered solids. It took researchers over a decade to discover certain deficiencies of the Miller–Abrahams resistor network model. Moreover, it was realized independently by Ambegaokar, Halperin, and Langer (1971), Shklovskii and Efros (1971), and Pollak (1972) that hopping conduction in semiconductors can be modeled successfully by using the concepts of percolation theory. Since their seminal papers, several electronic properties of semiconductors have been successfully predicted by percolation theory. This chapter summarizes the important elements of this successful application of percolation to a technologically important problem. Our discussion is by no means exhaustive as the number of published papers on the subject is too large. A thorough discussion can be found in the monograph of Shklovskii and Efros, (1984). An older account is given by Pollak (1978). We discuss the

Miller–Abrahams resistor network, then the application of percolation to predicting the hopping conductivity of disordered solids. This utilizes a modification of the Miller–Abrahams network model. We also discuss recent studies of fractal structure and hopping conductivity.

13.1 The Miller–Abrahams network

The starting point is the Boltzmann equation

$$\frac{\partial P_i}{\partial t} = \sum_j [W_{ji}P_j(1 - P_i) - W_{ij}P_i(1 - P_j)], \tag{13.1}$$

where P_i is the probability that site i is occupied. It is implicitly assumed that the occupation probabilities are uncorrelated. If repulsion can cause strong correlations, then the exclusion factor $(1 - P_i)$ should be omitted. In the linear (ohmic) regime, the current is proportional to the applied field, and we linearize (13.1) by writing

$$P_i = P_i^0 + \Delta P_i, \tag{13.2}$$

$$W_{ij} = W_{ij}^0 + \Delta W_{ij}, \tag{13.3}$$

where 0 denotes the equilibrium value, and Δ an increment proportional to the applied field. This linearization implies that $\Delta W_{ij} = -\Delta W_{ji}$. We thus obtain the linearized version of (13.1)

$$\frac{\partial \Delta P_i}{\partial t} + \sum_j A_{ij}\Delta P_i - \sum_j A_{ji}\Delta P_j = \sum_j B_{ji}\Delta W_{ji}, \tag{13.4}$$

where

$$A_{ij} = W_{ij}^0(1 - P_j^0) + W_{ji}P_j^0, \tag{13.5}$$

$$B_{ij} = P_i^0(1 - P_j^0) + P_j^0(1 - P_i^0). \tag{13.6}$$

Equation (13.4) is a set of linear equations for the unknown ΔP_i. The equilibrium values P_i^0 are given by the Fermi distribution

$$P_i^0 = \frac{1}{\exp(E_i/k_bT) + 1}, \tag{13.7}$$

where E_i is the energy of a carrier on site i, measured from the Fermi level, and k_b is the Boltzmann constant. The equilibrium values W_{ij}^0 are given by

$$W_{ij}^0 = \frac{u_{ij}}{|\exp[(E_j - E_i)/k_bT] - 1|}, \tag{13.8}$$

with

$$u_{ij} = u_{ji} = \frac{1}{\tau_0} \exp(-2r_{ij}/a). \tag{13.9}$$

In (13.9) $1/\tau_0$ is of the order of a phonon frequency, r_{ij} is the distance between i and j, and a is a Bohr radius. It is assumed that τ_0 depends only weakly on r_{ij} and T.

Suppose that $\mathbf{F}$ is the intensity of an applied field, and $\mathbf{r}_i$ is the radius vector of site i. The applied field changes the energy difference Δ_{ij} between the energies of sites i and j. Thus, in a linearized theory we should have

$$\Delta W_{ij} = \frac{dW_{ij}}{d\Delta_{ij}} e\mathbf{F} \cdot (\mathbf{r}_i - \mathbf{r}_j) = \frac{e\mathbf{F} \cdot (\mathbf{r}_i - \mathbf{r}_j)}{\sinh^2(\Delta_{ij}/k_b T)} u_{ij}, \tag{13.10}$$

where e is the charge of an electron. Miller and Abrahams defined a new variable V_i such that

$$P_i = P_i^0 + \Delta P_i \equiv \frac{1}{|\exp[(E_i - eV_i)/k_b T] + 1|}, \tag{13.11}$$

so that to first order

$$\Delta P_i = \frac{dP_i^0}{dE_i} eV_i = \frac{eV_i}{4k_b T \cosh^2(E_i/2k_b T)}, \tag{13.12}$$

In the linear regime the variable V_i is proportional to $\mathbf{F}$. We can transform the set of linear equations for P_i to another set for V_i. The resulting set of linear equations is then given by

$$D_i \frac{\partial V_i}{\partial t} = \sum_j D_{ji} V_j - \sum_j D_{ij} V_i + \sum_j G_{ij} \mathbf{F} \cdot \mathbf{r}_{ij}, \tag{13.13}$$

where $D_i = P_i^0(1 - P_i^0)$, $D_{ij} = D_i A_{ij}$, and $G_{ij} = B_{ij} W_{ij}^0 W_{ji}^0 / u_{ij}$.

We can now discuss the construction of a network model for calculating the hopping conductivity. Consider first the steady state. We define a temperature dependent conductance G_{ij} by

$$\frac{k_b T G_{ij}}{e^2} = P_i^0(1 - P_j^0) W_{ij}^0 = P_j^0(1 - P_i^0) W_{ji}^0. \tag{13.14}$$

If we substitute (13.14) into the steady state limit of (13.13), we obtain

$$\sum_j \left\{ \left[V_i - \frac{\mathbf{F} \cdot \mathbf{r}_i(W_{ij}^0 + W_{ji}^0)}{u_{ij}} \right] - \left[V_j - \frac{\mathbf{F} \cdot \mathbf{r}_j(W_{ij}^0 + W_{ji}^0)}{u_{ij}} \right] \right\} G_{ij} = 0, \tag{13.15}$$

where $(W_{ij}^0 + W_{ji}^0)/u_{ij} = \coth(|\Delta_{ij}|/2k_bT)$. We are mainly interested in the regime for which $\coth(|\Delta_{ij}|/2k_bT) \sim 1$, in which case (13.15) becomes

$$\sum_j [(V_i - \mathbf{F}\cdot\mathbf{r}_i) - (V_j - \mathbf{F}\cdot\mathbf{r}_j)]G_{ij} = 0. \tag{13.16}$$

Equation (13.16) represents a network of resistors if we think of $(V_i - \mathbf{F}\cdot\mathbf{r}_i)$ as the potential at site i. Then, $Z_{ij} = 1/G_{ij}$ is the resistance between sites i and j, and (13.16) is simply Kirchhoff's equation for site j. Miller and Abrahams treated Z_{ij} more generally and considered it as an impedance.

For the unsteady state, the time-dependent term of (13.13) does not vanish, and (13.16) must be rewritten as

$$\frac{P_i^0 e^2(1 - P_i^0)}{k_bT}\frac{\partial V_i}{\partial t} = \sum_j [(V_i - \mathbf{F}\cdot\mathbf{r}_i) - (V_j - \mathbf{F}\cdot\mathbf{r}_j)]G_{ij}. \tag{13.17}$$

To make a more general network for this case, we define a capacitance $C = P_i^0 e^2(1 - P_i^0)/k_bT$ with a potential V_i across it. We now refer all the potentials to "ground" potential, which is zero. Because $\mathbf{F}\cdot\mathbf{r}_i$ is the applied potential at i, it is represented as an output from a generator connected in series with C between the ground and site i. There is an impedance Z_{ij} connected between any two junctions i and j. There is also a capacitor C_i in series with a generator connected to the ground. Using the expression for P_i^0 and W_{ij}^0, and restricting our attention to the case where various site energies are of the order or larger than k_bT, we obtain

$$Z_{ij} = k_bT\frac{\exp[(|E_i| + |E_j| + |E_i - E_j|)/2k_bT]}{e^2 u_{ij}}, \tag{13.18}$$

$$C_i = \frac{e^2}{k_bT}\exp(-E_i/k_bT). \tag{13.19}$$

Using (13.9), we can rewrite (13.18) as

$$Z_{ij} = \frac{k_bT}{e^2}\exp(E_{ij}/k_bT + 2r_{ij}/a)\tau_0, \tag{13.20}$$

where E_{ij} is either the energy of the site farther from the Fermi energy, or $E_{ij} = (|E_i| + |E_j| + |E_i - E_j|)/2$. Equations (13.18) and (13.19) have an important implication. Even if the site energies are moderately distributed, the exponential dependence of Z_{ij} and C_i on these energies makes them enormously broadly distributed. This can be used to reduce the computations of the effective properties of the network, since the broadness of the distribution of Z_{ij} implies that there are many small conductances that can be removed from the network. The resulting network is called the *reduced* network.

The procedure for constructing the reduced network is discussed in great details by Pollak (1978). Here we give a summary of his discussion. As the first step, we select only the largest capacitances (for example, those within a given factor) that exist in the network. All such capacitances are then considered as equal with the common value C. All other capacitances and their associated sites are then deleted from the network. Next, a resistance $Z_\ell = 2/(\omega C)$ is determined, where ω is the frequency at which the properties of the network are to be calculated. We then discard all resistances that are larger than Z_ℓ, and replace all resistances that are smaller than Z_ℓ by shorts. At high frequencies Z_ℓ is very small, so there are very few resistances smaller than or equal to Z_ℓ. But as the frequency is lowered, clusters of connected resistors appear. These clusters merge together and form a sample-spanning cluster when a critical frequency ω_c is reached. This is similar to the formation of sample-spanning clusters in percolation networks. We have a critical resistance $Z_c = 2/(\omega_c C)$, at which a sample-spanning cluster is formed for the first time. If we use a frequency smaller than ω_c, the reduction procedure becomes ineffective; replacing the resistances by shorts produces a macroscopic short throughout the network.

Miller and Abrahams were the first to calculate the hopping conductivity G of semiconductors using reduced networks. They assumed that the statistical distribution of the resistances depends only on r_{ij} and not the site energies. This was justified because the experimental data for some semiconductors indicated that impurity conduction exhibits a well-defined activation energy. But Mott (1968) pointed out that the exponential dependence of the resistances on the site energies cannot be ignored in most cases; if the activation energy of a nearest-neighbor site is large, a hop to a distant site whose energy is lower may be easier than one to a nearest-neighbor site. How far the hopper can go depends on the ease of activation to higher energies, therefore the hopping distance, and thus the resistance, depends on the temperature. This mechanism of hopping conduction is usually called *variable range hopping*. It contrasts with the original work of Miller and Abrahams, which was restricted to nearest-neighbor hopping. It is now generally believed that variable range hopping is the appropriate mechanism at low temperatures, whereas nearest-neighbor hopping may be appropriate at high temperatures. At low temperatures, Mott showed that

$$G = G_0 \exp[-(T_0/T)^\alpha], \tag{13.21}$$

an important characteristic of variable range hopping conductivity. It is now one of most famous results for hopping conductivity of semiconductors. In general, α depends on the density of states near the Fermi level. In Mott's theory the density of states (see Chapter 10) was assumed to be constant, which results in $\alpha = 1/(d+1)$ for a d-dimensional system, G_0 and T_0 are some constants, and

$$T_0 = \frac{\lambda a^3}{k_b N}, \tag{13.22}$$

where N is the density of states (see Chapter 10) at the Fermi level E_F, and λ is a dimensionless parameter. Equation (13.21) is particularly accurate for amorphous Ge in the range $60\text{K} \leqslant T \leqslant 300\text{K}$, with $T_0 \simeq 7 \times 10^7\text{K}$. Similar temperature dependences have also been found in amorphous silicon and carbon, and in vanadium oxide. According to Pollak (1978), unless G is measured over several orders of magnitudes, a $T^{-1/4}$ behavior should be treated with caution and should not automatically be interpreted as evidence for variable range hopping conductivity. Hill (1976) analyzed most of the published experimental data and showed that most of them do follow (13.21) with $\alpha = 1/4$. We discuss shortly the conditions under which G might deviate from $\alpha = 1/4$.

Over a decade after the original work of Miller and Abrahams (1960), Ambegaokar *et al.* (1971), Shklovskii and Efros (1971), Pollak (1972), and Brenig *et al.* (1971) reexamined the transport paths. They realized the paths that Miller and Abrahams had thought to be carrying most of the current in the network do not in fact carry any current in most situations. The first three groups used percolation to find the correct current-carrying paths. If we always proceed through nearest-neighbors as in the Miller–Abrahams theory, we are certain to arrive at a site where our nearest-neighbor is a large distance away, so it may be more efficient to go through nonnearest-neighbors. That is why the Miller–Abrahams paths do not usually carry current. We now discuss the percolation models for calculating the hopping conductivity of semiconductors. We follow Ambegaokar *et al.* (1971) whose work is very elegant and conceptually simple.

13.2 Percolation models of hopping conductivity

Ambegaokar *et al.* argued that an accurate estimate of G is the critical percolation conductance G_c, which is the largest value of the conductance such that the subset of the network with $G_{ij} > G_c$ still contains a conducting sample-spanning cluster. They divided the network into three parts. (1) A set of isolated clusters of high conductivity; each cluster consists of a group of sites connected together by conductances $G_{ij} >> G_c$. (2) A small number of resistors with G_{ij} of order G_c, which connect together a subset of high conductance clusters to form the sample-spanning cluster. This was called *the critical subnetwork*, essentially the same as the static limit of the reduced network discussed above. (3) The remaining resistors with $G_{ij} << G_c$. The resistors in (2) dominate the overall conductance of the network. The same ideas were used by Katz and Thompson for estimating the permeability of a porous medium with a broad pore size distribution; see Chapter 5. The condition that $G_{ij} > G_c$, together with (13.20), can be expressed as

$$\frac{r_{ij}}{r_m} + \frac{|E_i| + |E_j| + |E_i - E_j|}{2E_m} \leqslant 1, \tag{13.23}$$

where $r_m = a\ln(G_0/G_c)/2$ is the maximum distance between any two sites between which a hop can occur, and $E_m = k_bT\ln(G_0/G_c)$ is the maximum energy that any initial or final state can have.

To construct the critical subnetwork, Ambegaokar *et al.* considered an empty network and, starting with the smallest resistors, inserted them in the network one by one. As more resistors are inserted, clusters of connected resistors are formed until the critical value $Z_c = 1/G_c$ is reached at which a sample-spanning cluster is formed. To calculate Z_c they used (13.20) and assumed that τ_0 is constant. Moreover, they also assumed that the density of states N is constant near the Fermi level, $N(E) = N(E_F)$. Thus, Z_{ij} is a monotonic function of the random variable $\zeta = E_{ij}/k_bT + 2r_{ij}/a$, and the critical value Z_c defines a corresponding critical value ζ_c. Around each site i such that $E_i < \zeta k_bT/2$, a sphere of radius

$$r_i = \frac{a}{2}\left(\frac{\zeta}{2} - \frac{E_i}{k_bT}\right), \tag{13.24}$$

is drawn. The radius of the sphere increases with ζ. When two spheres overlap, a bond is inserted between the two sites with overlapping spheres. This happens only if (13.23) is satisfied. Percolation occurs at ζ_c, corresponding to a critical radius r_c. Ambegaoker *et al.* formulated this problem as a site percolation process. Pollak (1972) treated it as a bond percolation phenomenon; this allowed him to include the effect of short-range correlations but his basic results are the same as those of Ambegaokar *et al.* Equation (13.24) tells us that there is a maximum radius $r_m = a\zeta_c/4$ and a maximum energy $E_m = \zeta_c k_bT/2$, with $\zeta_c = 2\ln(G_0/G_c)$. If two sites are separated by a distance larger than r_m, or farther from energy level E_m, they will not contribute significantly to the hopping conductivity. The volume of the sphere defined by (13.24) is $(\pi/6)a^3(\zeta_c/2 - E_i/k_bT)^3$, and the volume averaged over all sites with a sphere of nonzero radius (i.e., those for which $E_i < \zeta k_bT/2$) is $\langle V\rangle$ given by

$$\langle V\rangle = \frac{\pi}{384}a^3\zeta_c^3 = \frac{\pi}{48}r_m^3. \tag{13.25}$$

The volume fraction ϕ_c of the spheres at the percolation threshold, is given by $\phi_c \simeq n\langle V\rangle$, where, assuming that the density of states is constant, $n = N\zeta_c k_bT$ is the fraction of the sites with a sphere, i.e., those with an energy in the interval $(-E_m, E_m)$. Ambegaokar *et al.* estimated that $\phi_c \simeq 1/4$, somewhat larger than $\phi_c \simeq 0.15$–0.17, estimated by Scher and Zallen (1970; see Chapter 2) for three-dimensional percolating continua. We can also calculate the number of bonds per site B_c of the network. We learned in Chapter 2 that for bond percolation $B_c \simeq d/(d-1)$. For three-dimensional amorphous materials (or continuum percolation), $B_c = 4\pi n r_c^3/3$, and computer simulations of Pike and Seager (1974) indicated that $B_c \simeq 2.8$ for three-dimensional systems. On the other hand, B_c is related to the density of states $N(E)$ by

$$B_c = \frac{4\pi n}{3} \frac{\int_{-E_m}^{+E_m} N(E_i)dE_i \int_{-E_m}^{+E_m} (r_m^3 + 3r_m^2 D_m + 3r_m D_m^2) N(E_j)dE_j}{\int_{-E_m}^{+E_m} N(E_i)dE_i}, \tag{13.26}$$

where D_m is the mean size of the sites. Combining $\phi_c = n\langle V\rangle \simeq 1/4$ with (13.25), or using (13.26) together with the lattice or continuum value of B_c and the appropriate expression for r_m, we finally obtain

$$G_c = G_0 \exp[-(T_0/T)^{1/4}], \tag{13.27}$$

$$T_0 = \frac{16a^3}{k_b N}. \tag{13.28}$$

Comparing (13.28) with (13.22) indicates that the percolation model of Ambegaokar *et al.* predicts that $\lambda \simeq 16$. Their model not only predicts the $T^{-1/4}$ behavior proposed by Mott, it also provides an estimate of the temperature T_0 defined by (13.21).

The preexponential factor G_0 has also been calculated by several research groups, since quantitative prediction of G_c requires an accurate value of G_0. For example, using some of the ideas of Kurkijärvi (1974), Shklovskii and Efros (1975) proposed that $Z_0 = 1/G_0 = r_c(2r_c/a)^{\nu_p} R_0$, where ν_p is the correlation length exponent of three-dimensional percolation, and R_0 is the resistance for $\zeta = \zeta_c$. Kirkpatrick (1974) suggested the same expression, except that in his equation ν_p is replaced by $(\mu - 1)$, where μ is the critical exponent of the conductivity of three-dimensional percolation. Pollak (1972), Butcher and McInnes (1978), Butcher (1980), Movaghar *et al.* (1980a, b), and Movaghar and Schirmacher (1981) also calculated the preexponential factor G_0, although their results did not involve any critical exponent of percolation. The predictions of Butcher, and Movaghar and coworkers are particularly accurate.

Equation (13.27) gives a lower bound to the true hopping conductivity of a network whose individual conductances vary over a broad range. That (13.27) is a lower bound to the true G is because the critical subnetwork corresponds to replacing all $G_{ij} < G_c$ by 0, and all $G_{ij} \geqslant G_c$ by G_c in the original network. Equation (13.27) is exact only in the limit $T \to 0$. If $T > 0$, hops with conductance less than G_c also contribute to the macroscopic conductivity; this means that the optimal cutoff should be somewhat larger than G_c. Moreover, the percolation approach of Ambegaokar *et al.* cannot be used for one-dimensional, or quasi-one-dimensional conductors, since percolation disorder divides a linear chain into finite segments and the problem becomes meaningless. The physical systems to which this situation may be relevant are two classes of compounds that consist of weakly coupled parallel chains of strongly

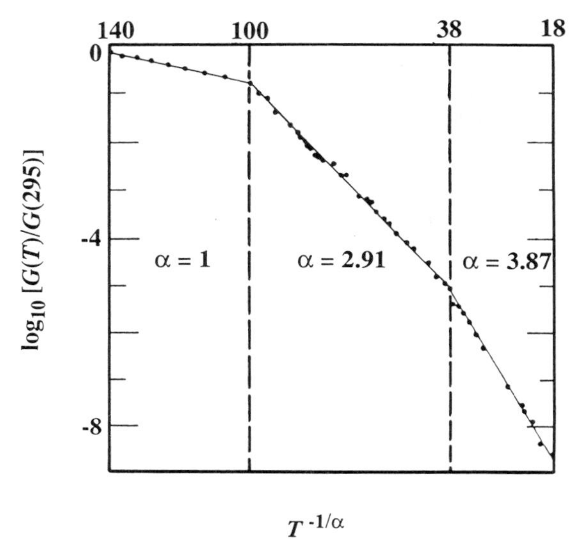

Figure 13.1 *Temperature dependence of the DC conductivity of NMP–TCNQ. T is in kelvin, and the straight lines are the best fit to the data (after Shante 1977).*

coupled atoms or molecules. Their conductivity is highly anisotropic, therefore they may be treated as essentially one-dimensional conductors. Well-known examples are salts of the organic ion-radical tetracyanoquinodimethane (TCNQ) and the square planar complexes of transition metals such as platinum and iridium. Shante (1977) proposed a modification of the percolation model of Ambegaokar *et al.* that appears to take into account the effect of such complexities. The model is a bundle of chains in which hopping can occur along the chains and between them. Interchain hopping was assumed to be much more difficult than intrachain hopping, but it was still allowed. Shante's model also allowed the possibility of intrachain hoppings in either one or two dimensions. At low temperatures Shante's model corresponded to two- and three-dimensional percolation, respectively, consequently $T^{-1/3}$ and $T^{-1/4}$ behaviors were obtained. At high temperatures the percolation model is no longer applicable, and Shante obtained a T^{-1} behavior. Figure 13.1 compares the predictions of his model with the experimental data for *N*-methylphenazinium-TCNQ compounds, and the agreement is excellent.

13.3 Effect of a variable density of states on hopping conductivity

In the percolation model discussed above, we assumed the density of states is constant near the Fermi level. Although the basic $T^{-1/4}$ law has been observed in many systems (see, e.g., Knotek *et al.* 1973, Viščor and Yoffe

1982, among others), criticism of this formulation was raised by many (see, e.g., Szpilka and Visčor 1982), mainly because the predicted and measured values of G_0 differed by several orders of magnitude. It has been suggested that the assumption of a constant density of states in the above percolation model may not be justified; instead we have to use a variable density of states, i.e., (13.26) with a variable density of states.

Ortuno and Pollak (1983) investigated this problem and proposed that, if the density of states is concave, then an appropriately modified percolation model can explain the data and remove the disagreement between the predictions and the experimental data. They used (13.26) with $N(E) = N(E_F)\exp(E/E_0)$, where E_0 is the exponential decay rate and treated the problem in details. They compared the predictions with the experimental data for amorphous Si and amorphous Ge, and found good agreement between the predictions and the data. Moreover, the predicted value of G_0 was of the order of the experimental data. Maloufi *et al.* (1988) used the theory of Ortuno and Pollak to fit their conductivity data for amorphous Si_ySn_{1-y} with $y = 0.47$–1. Figure 13.2 shows the fits of their data by this theory, and the agreement appears to be excellent over much of the temperature range.

As another example, consider the case for which $N(E) \sim |E|^{\beta}$, where β is a positive constant. This case was treated by Hamilton (1972) and Pollak (1972). Using (13.26) we obtain (13.21) with $\alpha = (\beta + 1)/(\beta + 4)$, which reduces to (13.21) when $\beta = 0$. The limit $\beta \to \infty$ is also interesting because

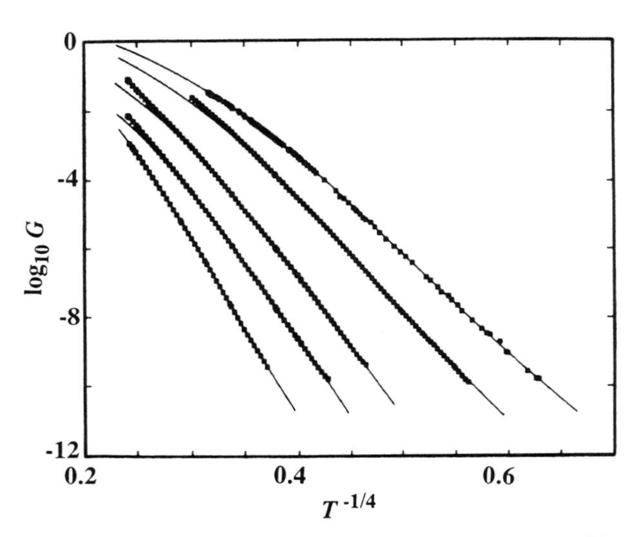

Figure 13.2 *Comparison of the predicted temperature dependence of hopping conductivity G of a-Si_ySn_{1-y} (straight lines), obtained with an exponential density of states, with the experimental data (symbols). The data are, from right to left, for* $y = 0.47$, 0.62, 0.77, 0.9, *and 1, and T is in kelvin (after Maloufi et al. 1988).*

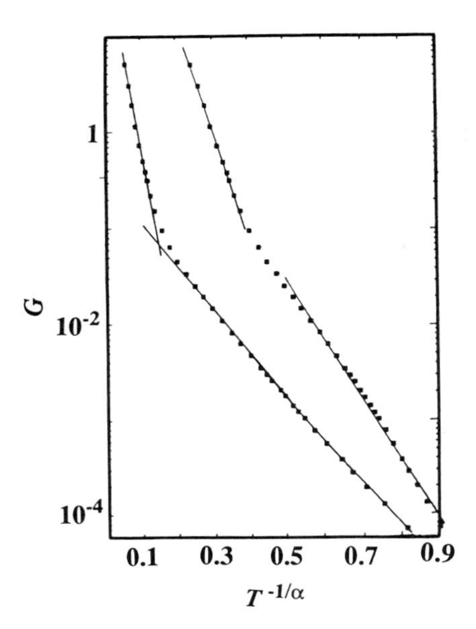

Figure 13.3 *Temperature dependence of hopping conductivity G of doped and compensated GaAs. Symbols are experimental data. The straight lines on the right represent* $T^{-1/4}$ *fits, while those on the left represent* $T^{-1/2}$ *fits (after Redfield 1973).*

it corresponds to a system that has a sudden onset of states away from the Fermi level; we thus obtain a T^{-1} behavior. An experimental realization of a power law density of states was provided by Redfield (1973), who carried out a careful study of hopping conductivity of heavily doped and strongly compensated GaAs. His results indicated a $T^{-1/2}$ behavior rather than $T^{-1/4}$. Redfield showed that, although $T^{-1/4}$ might look plausible, his data could be fitted extremely well by a $T^{-1/2}$ law; see Fig. 13.3. His data can be easily interpreted if we take $\beta = 2$.

Sheng and Klafter (1983) studied hopping conductivity of disordered granular media, where conduction results from tunneling of electrons and holes from charged grains to neutral ones. Electrons have to be transferred from one neutral grain to another. This requires each grain to be characterized by a charging energy $E_c = e^2/(\varepsilon D)$, where ε is the dielectric constant and D is the grain size. A disordered granular medium is thus characterized by a distribution $f(E_c)$, related to the grain size distribution, and the density of states is related to this distribution by

$$N(E) = \frac{1}{\langle \Delta E \rangle} \int_0^E f(E_c) dE_c, \tag{13.29}$$

where $\langle \Delta E \rangle$ is the average electronic level separation inside conducting grains. Thus, a given distribution of energies $f(E_c)$ can be immediately translated into one for the density of states. Normally, we expect that $N(0) = 0$. But in any composite material we can have energy states other than those in the conducting grains. That is, we can have impurities that could contribute a finite $N(E)$ at $E = 0$; there is some experimental evidence for this. Thus, Sheng and Klafter (1983) assumed that $N(E) = N_0 + 1/\langle \Delta E \rangle$ and a log-normal distribution for $f(E_c)$, where $N_0 = N(E_F)$. Then they calculated the hopping conductivity of a granular medium using (13.26). In their two-parameter fit of the results, the fitting parameters were the width of the distribution $f(E)$ and $x = \pi n D_m/6$. By varying these two parameters they could obtain a variety of conductivity behavior, ranging from $T^{-1/4}$ to $T^{-1/2}$. If, instead of the density of states that Sheng and Klafter used, we use $N(E) = N_0 + cE^{\beta}$, where c is a constant, we obtain (13.21) with $\alpha = (\beta + 1)/(\beta + 2)$. This may correspond to a system with a broad particle size distribution; Mehbod *et al.* (1987) showed how it fits very well with their experimental data for hopping conductivity of polymer-conducting carbon black composites. Randomly dispersed in the matrix are particles that have a broad size distribution. The polymeric matrices were polystyrene, polyethylene, ethylene–propylene · copolymer, and styrene–butadiene copolymer.

13.4 Effect of Coulomb interactions on hopping conductivity

The $T^{-1/2}$ behavior of hopping conductivity has been observed in several systems. Although, a variable density of states, such as $N(E) \sim E^{\beta}$ or $N(E) = N_0 + cE^{\beta}$, can explain the data, the origin of these power laws was not clear for some time. An explanation for them was proposed by Efros and Shklovskii (1975). They suggested that Coulomb interactions between localized electrons (which are long-range interactions) can create a soft gap (called a Coulomb gap) in the density of states near the Fermi level. This means that in a narrow gap centered around the Fermi level, the density of states cannot be constant, and has to vary with the energy, whereas outside the gap the density of states vanishes. If δ is the width of this gap, Efros and Shklovskii (1975) showed that $N(\delta) \sim \delta^2$ in three dimensions and $N(\delta) \sim \delta$ in two dimensions. As a result, they suggested that in three dimensions there should be a crossover from $T^{-1/4}$ behavior at relatively high temperatures, where the Coulomb gap is not effective, to $T^{-1/2}$ behavior at lower temperatures. Although $T^{-1/2}$ behavior had been reported by several groups, the temperature below which the gap could be detected by conductivity measurements is usually too low in amorphous semiconductors. Generally speaking, the Coulomb gap does not affect the conductivity of amorphous semiconductors. Moreover, the Coulomb gap cannot be found in good metals. It can only affect systems that have localized electronic states. A Coulomb gap affects the conductivity of doped crystallic semiconductors, and experimental evidence for the crossover from $T^{-1/4}$ to $T^{-1/2}$ behavior was provided by these systems.

The first convincing experimental evidence for this crossover was probably provided by Zhang *et al.* (1990). They measured resistivity of five insulating compensated *n*-type CdSe samples – doped semiconductors. Their data clearly indicated a crossover from $T^{-1/4}$ behavior to $T^{-1/2}$ as the temperature was decreased by about three orders of magnitude from 15 to 0.04K. Over this temperature range the hopping energy becomes comparable and then smaller than the energy gap, discussed by Efros and Shklovskii (1975). The crossover temperature T_{co} was found to decrease as the concentration of the donor was increased; this is expected. Aharony *et al.* (1992) proposed a scaling theory for this crossover. They proposed that the resistivity $Z = 1/G$ of the sample obeys the following scaling law

$$\ln(Z/Z_0) = Ah(T/T_{co}), \tag{13.30}$$

where the scale factor A and the crossover temperature T_{co} depend on the sample properties, but the scaling function $h(x)$ is *universal* and has the limiting behaviors, $h(x) \sim x^{-1/4}$ for $x >> 1$ and $h(x) \sim x^{-1/2}$ for $x << 1$. Moreover, using the percolation model, Aharony *et al.* proposed that

$$h(x) = \frac{x + [\sqrt{x+1} - 1]}{x[\sqrt{1+x} - 1]^{1/2}}, \tag{13.31}$$

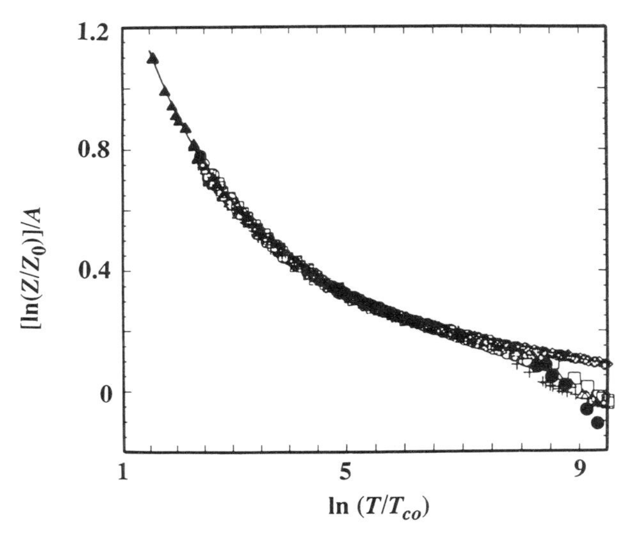

Figure 13.4 *Temperature dependence of hopping conductivity G of carbon black–polymer composites. T is in kelvin. The slope of the line is* α ≃ 2/3 *(after van der Putten et al. 1992).*

and provided expressions for A and the crossover temperature T_{co}. The implication of (13.30) and (13.31) is that, if $\ln(Z/Z_0)$ for various samples are plotted as a function of T/T_{co}, all the data should collapse onto a single curve. Figure 13.4 shows such a collapse for the data of Zhang *et al.*, and the agreement is excellent. This agreement lends strong support to the scaling hypothesis of Aharony *et al.* and the validity of (13.30) and (13.31).

13.5 Effect of a fractal structure on hopping conductivity

A $T^{-1/2}$ behavior for hopping conductivity can be explained in terms of the Coulomb interactions and gap. But the Coulomb interactions are believed not to play any role in many semiconductors which do exhibit a $T^{-1/2}$ behavior; we need a different theoretical explanation for them. Although a variable density of states has been proposed as the source for this behavior, there has recently been some experimental evidence that one reason for the $T^{-1/2}$ behavior can be a fractal structure for the semiconductor. Deutscher, Lévy, and Souillard (1987), using ideas of Lévy and Souillard (1987), proposed a new mechanism that may give rise to a conductivity behavior almost like $T^{-1/2}$. Lévy and Souillard (1987) suggested that, if a system is a fractal object, the impurity quantum states are *superlocalized*, i.e., their wave functions $\psi(r)$ decay with distance r *faster* than exponentially, $\psi(r) \sim \exp(-r^{\gamma})$ with $\gamma > 1$ (in the classical Anderson localization, $\gamma = 1$). Experimental evidence for superlocalization was provided by Tsujimi *et al.* (1988). They carried out very low-frequency depolarized Raman scattering measurements on fractal silica aerogels. If $I(\omega)$ is the total scattering intensity at frequency ω, it can be shown that

$$\frac{I(\omega)}{N(\omega)} \sim \omega^{-2(1-d_s\gamma/D_f)}, \tag{13.32}$$

where D_f is the fractal dimension of the system. Here $d_s = 2D_f/d_w$ is the spectral or fracton dimension discussed in Chapter 10, where d_w is the fractal dimension of a random walker in anomalous diffusion in the fractal system, discussed in Chapter 9. Measurements of Tsujimi *et al.* yielded $\gamma > 1$, indicating superlocalization. Lévy and Souillard also argued that for percolation clusters at the percolation threshold p_c, or, if the system is above p_c, for length scales smaller than the percolation correlation length ξ_p, we have $\gamma = d_w/2$, see (9.5). This yields $\gamma \simeq 1.43$ and 1.9 for two and three dimensions, respectively.

Based on the arguments of Lévy and Souillard, Deutscher *et al.* suggested that, as p_c is approached, the behavior of hopping conductivity should follow (13.21) but with

$$\alpha = \frac{\gamma}{\gamma + D_c}, \tag{13.33}$$

in which they used $\gamma = d_w/2$, where D_c is the fractal dimension of the largest percolation cluster at p_c (see Chapter 2). More generally, D_c may be replaced by the fractal dimension D_f of any fractal object whose random walk fractal dimension is d_w, and thus the potential applicability of (9.33) is not restricted to percolation systems. Chapter 10 shows that the spectral dimension for percolation systems $d_s = 2D_c/d_w$ is essentially constant and $d_s \simeq 4/3$, for both $d = 2$ and 3. If we assume $\gamma = d_w/2$ and use (13.33), we obtain $\alpha \simeq 3/7 \simeq 0.43$ for both $d = 2$ and 3. Based on this result, Deutscher *et al.* argued that the observed $T^{-1/2}$ behavior of hopping conductivity may in fact be a $T^{-3/7}$ behavior caused by the fractal structure of the system. They discussed the conditions under which their prediction may be observed in experiments.

However, Harris and Aharony (1987) and Aharony and Harris (1990) argued that (13.33) is incorrect. They argued that we have to distinguish between the behavior of a system with a *typical* geometry and the behavior of a system averaged over *all* the possible geometries. For typical geometries, they proposed that

$$\max(\nu_p^{-1}, 1) \leqslant \gamma \leqslant D_{min}, \tag{13.34}$$

where ν_p is the critical exponent of percolation correlation length, and D_{min} is the fractal dimension of the minimum or chemical path, discussed in Chapter 2. Using the numerical values of ν_p and D_{min}, we obtain $1 \leqslant \gamma(d = 2) \leqslant 1.13$ and $1.14 \leqslant \gamma(d = 3) \leqslant 1.34$. Hence the proposal of Lévy and Souillard that $\gamma = d_w/2$ cannot be correct. Harris and Aharony also showed that an average over all the possible geometries yields $\gamma = 1$. Lambert and Hughes (1991) used very accurate numerical simulations to obtain $\gamma \simeq 1.13 \pm 0.06$ and 1.39 ± 0.07 for two- and three-dimensional percolation networks, respectively. Thus, the upper bound of Harris and Aharony may in fact be an exact result, i.e., $\gamma = D_{min}$. Earlier, and presumably less accurate, simulations of de Vries *et al.* (1989) had yielded $\gamma(d = 2) \simeq 1.0$, i.e., we do *not* have superlocalization in two dimensions. Harris and Aharony also suggested that for hopping conductivity of a percolation system in the fractal regime

$$\alpha \simeq \frac{D_{min}}{D_c + D_{min}}, \tag{13.35}$$

consistent with $\gamma = D_{min}$. Equation (13.35) then yields $\alpha = 0.37$ and 0.35 for $d = 2$ and 3, respectively, indicating that the existence of a fractal structure *cannot* by itself explain the observed $T^{-1/2}$ behavior of hopping conductivity of semiconductors in which Coulomb interactions are unimportant.

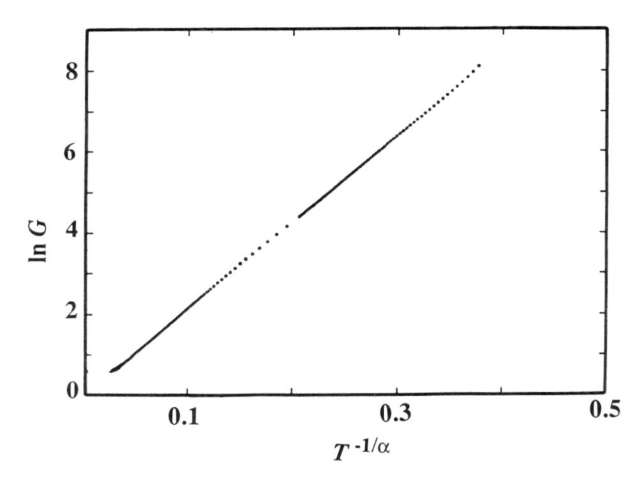

Figure 13.5 *Temperature-dependence of the hopping conductivity G of carbon black–polymer components (after van der Putten et al. 1992).*

But (13.35) has not yet been confirmed by any experiment. van der Putten *et al.* (1992) measured the conductivity of carbon black–polymer composites as a function of carbon black concentration from a point in the vicinity of the percolation threshold up to $33p_c$, and in the temperature range 4–300 K. Their data are shown in Fig. 13.5, from which we obtain $\alpha \simeq 2/3$. This does not agree with the prediction of (13.35). To interpret this result, they assumed a power law density of states, $N(E) \sim E^{\beta}$, and modified the result of Deutscher *et al.* to obtain

$$\alpha = \frac{\gamma(\beta + 1)}{D_c + \gamma(\beta + 1)}. \tag{13.36}$$

If we now use $\alpha = 2/3$, $D_c \simeq 2.52$, and $\gamma \simeq D_{min} \simeq 1.34$, we obtain $\beta \simeq 11/4$. Their data can be best explained by a combination of a percolating fractal structure *and* a power law density of states. Moreover, if we take $\beta = 1$, then (13.36) predicts that $\alpha \simeq 1/2$. This means that a linear density of states *and* a percolating fractal structure may be responsible for the observed $T^{-1/2}$ behavior of hopping conductivity of semiconductors when Coulomb interactions may be unimportant.

13.6 Conclusions

Hopping conductivity of semiconductors can be predicted quantitatively by percolation theory. The results of this chapter also show that fractal structures formed near the percolation threshold may at least be partly responsible for the experimentally observed deviations from Mott's law (13.21) with $\alpha = 1/4$.

Other properties of hopping conduction in semiconductors can also be predicted by percolation theory. The reader is referred to the monograph of Shklovskii and Efros (1984) for a detailed discussion.

References

Aharony, A. and Harris, A.B., 1990, *Physica* **A163**, 38.
Aharony, A., Zhang, Y., and Sarachik, M.P., 1992, *Phys. Rev. Lett.* **68**, 3900.
Ambegaokar, V., Halperin, B.I., and Langer, J.S., 1971, *Phys. Rev. B* **4**, 2612.
Brenig, W., Döhler, G., and Wölfle, 1971, *Z. Phys.* **246**, 1.
Butcher, P.N., 1980, *Phil. Mag.* **B42**, 799.
Butcher, P.N. and McInnes, J.A., 1978, *Phil. Mag.* **32**, 249.
Conwell, E.M., 1956, *Phys. Rev.* **103**, 51.
de Vries, P., De Raedt, H., and Lagendijk, A., 1989, *Phys. Rev. Lett.* **62**, 2515.
Deutscher, G., Lévy, Y., and Souillard, B., 1987, *Europhys. Lett.* **4**, 577.
Efros, A.L. and Shklovskii, B.I., 1975, *J. Phys. C* **8**, L49.
Hamilton, E.M., 1972, *Phil. Mag.* **26**, 1043.
Harris, A.B. and Aharony, A., 1987, *Europhys. Lett.* **4**, 1355.
Hill, R.M., 1976, *Phys. Stat. Sol.* **A35**, K29.
Hung, C.S. and Gliessman, J.R., 1950, *Phys. Rev.* **79**, 726.
Knotek, M.L., Pollak, M., Donovan, T.M., and Kurtzman, H., 1973, *Phys. Rev. Lett.* **30**, 853.
Kirkpatrick, S., 1974, in *Proceedings of 5th International Conference on Amorphous and Liquid Semiconductor*, London, Taylor and Francis, p. 183.
Kurkijärvi, J., 1974, *Phys. Rev. B* **9**, 770.
Lambert, C.J. and Hughes, G.D., 1991, *Phys. Rev. Lett.* **66**, 1074.
Lévy, Y.-E. and Souillard, B., 1987, *Europhys. Lett.* **4**, 233.
Maloufi, N., Audouard, A., Piecuch, M., Vergant, M., Marchal, G., and Gerl, M., 1988, *Phys. Rev. B* **37**, 8867.
Mehbod, M., Wyder, P., Deltour, R., Pierre, C., and Geuskens, G., 1987, *Phys. Rev. B* **36**, 7627.
Miller, A. and Abrahams, E., 1960, *Phys. Rev.* **120**, 745.
Mott, N.F., 1956, *Can. J. Phys.* **34**, 1356.
Mott, N.F., 1968, *J. Non-Cryst. Solids* **8–10**, 1.
Movaghar, B., Pohlmann, B., and Schirmacher, W., 1980a, *Phil. Mag.* **B41**, 49.
Movaghar, B., Pohlmann, B., and Schirmacher, W., 1980b, *Solid State Commun.* **34**, 451.
Movaghar, B. and Schirmacher, 1981, *J. Phys. C* **14**, 859.
Ortuno, M. and Pollak, M., 1983, *Phil. Mag.* **47**, L93.
Pike, G.E. and Seager, C.H., 1974, *Phys. Rev. B* **10**, 1421.
Pollak, M., 1972, *J. Non-Cryst. Solids* **11**, 1.
Pollak, M., 1978, in *The Metal–Non-metal Transition in Disordered Systems*, edited by L.R. Friedman and D.P. Tunstall, University of Edinburgh, p. 95.
Redfield, D., 1973, *Phys. Rev. Lett.* **30**, 1319.
Shante, V.K.S., 1977, *Phys. Rev. B* **16**, 2597.
Sheng, P. and Klafter, J., 1983, *Phys. Rev. B* **27**, 2583.
Shklovskii, B.I. and Efros, A.L., 1971, *Sov. Phys.-JETP* **33**, 468.
Shklovskii, B.I. and Efros, A.L., 1975, *Sov. Phys.-Tech. Phys. Lett.* **1**, 83.

Shklovskii, B.I. and Efros, A.L., 1984, *Electronic Properties of Doped Semiconductors*, Berlin, Springer.
Szpilka, A.M. and Visčor, P., 1982, *Phil. Mag.* **45**, 485.
Tsujimi, Y., Courtens, E., Pelous, J., and Vacher, R., 1988, *Phys. Rev. Lett.* **60**, 2757.
van der Putten, D., Moonen, J.T., Brom, H.B., Brokken-Zijp, J.C.M., and Michels, M.A.J., 1992, *Phys. Rev. Lett.* **69**, 494.
Visčor, P. and Yoffe, A.D., 1982, *J. Non-Cryst. Solids* **35/36**, 409.
Zhang, Y., Dai, P., Levy, M., and Sarachik, M.P., 1990, *Phys. Rev. Lett.* **64**, 2687.

14
Percolation in biological systems

14.0 Introduction

Most biological systems are so complex they preclude any reasonable description of them in terms of the basic interactions among their fundamental constituents. For this reason application of statistical physics of disordered systems, in particular percolation theory, to biological problems has thus far been relatively limited. But there are biological processes that are particularly statistical in nature and in which the role of connectivity of different elements or constituents is prominent. Examples include self-assembly of tobacco mosaic and other simple viruses (see, e.g., Hohn and Hohn 1970), actin filaments (Poglazov *et al.* 1967) and flagella (Asakura *et al.* 1968), lymphocyte patch and cap formation (Karnovsky *et al.* 1972), and many precipitation and agglutination phenomena. Some of these phenomena, such as precipitation, occur spontaneously if the functional groups are sufficiently reactive. Thus, these phenomena depend on their level of chemical complexity and that of the solvent in which they occur. Other factors are not directly related to the solvent but have great influence on the outcome of biological processes. For example, in antigen–antibody reactions, clusters of all sizes react with one another forming complex branched networks, which grow in size as time progresses. We may also have reactions which can proceed by rapid addition of monomers to growing chains after a slow initiating event. These processes are therefore similar to percolation processes. This chapter discusses application of percolation concepts to such phenomena. We start with a discussion of antigen–antibody reactions and aggregations, then discuss a few other biological processes to which percolation may be relevant.

14.1 Antigen–antibody reactions and aggregation

Under suitable conditions, mixtures of antigen and antibodies – systems comprised of bifunctional and multifunctional monomers – form large networks or aggregates that contain many antigen and antibody molecules. In general, these networks or branched structures react reversibly with one another. They are often insoluble and hence will precipitate. Since we are interested in the concentrations of antibodies and antigen as a function of time, it is important to examine the conditions under which precipitation may be expected to occur.

The first theory of antigen–antibody aggregation was developed by Goldberg (1952), who used combinatorial methods to derive results that are equivalent to those obtained for various percolation properties on a Bethe lattice. His work was extended by several authors including Aladjem and Palmiter (1965), Bell (1971), and Delisi (1974). Delisi and Perelson (1976) treated the problem extensively and derived several analytical results for various properties of interest.

Consider the aggregation of two types of monomers, S and L, of functionality 2 and $Z \geqslant 2$, respectively. The S monomers can be any of a number of substrates, e.g., IgG, while the L monomers might be some appropriate ligand or antigen. The functional units on these monomers are equivalent to the sites in percolation on random networks discussed so far. Three basic assumptions were made by Goldberg (1952): (1) all free sites are equivalent; (2) no intramolecular reactions can occur; and (3) during the aggregation process cyclization reactions do not occur. Closed loops are not formed; this immediately points to the similarity with percolation on Bethe lattices. Delisi and Perelson (1976) relaxed the second assumption to allow for intramolecular reactions. In the immunological literature such reactions are often referred to as monogamous bivalency.

Several parameters are introduced: (1) p_b, the probability that an *antibody* site picked at random is bound; (2) p_g, the probability that an *antigen* site picked at random is bound; (3) p_{br}, the probability that an antibody site is bound and has not reacted intramolecularly; and (4) p_{gr}, the probability that an antigen site is bound and has not reacted intramolecularly. Consider now an aggregate composed of x Z-functional antigens and $x - 1 + n$ bifunctional antibodies, where n is the number of antibodies that do not serve as connectors. Of these antibodies, n_1 are bound univalently and n_2 bivalently. An aggregate with compositions x, n_1, n_2 is called an (x, n_1, n_2)-mer. The main quantity of interest is $C_{n_1,n_2}(a, p)$, the concentration of such aggregates. To find this quantity, we define $P(x, n_1, n_2)$ as the probability that a free site picked at random is part of an (x, n_1, n_2)-mer. If closed loops are not formed, then

$$P(x, n_1, n_2) = \frac{\text{The number of free sites on } (x, n_1, n_2)\text{-mers}}{\text{Total number of free sites}}, \quad (14.1)$$

which means that

$$P(x, n_1, n_2) \equiv \frac{(Zx - 2x + 2 - 2n_2)C_{n_1,n_2}(x, p)}{L + S}, \tag{14.2}$$

where L and S are the number of the two types of monomers. The probability $P(x, n_1, n_2)$ can be written as

$$P(x, n_1, n_2) = \rho_l \omega_l \Omega_l + \rho_s \omega_s \Omega_s, \tag{14.3}$$

where l and s label free antigen and antibody sites respectively. Here ρ_l is the probability that the root is a free antigen site, ω_l the probability that the root is on an (x, n_1, n_2)-mer given that it is a free L site, and Ω_l the number of ways an (x, n_1, n_2)-mer can form given that the root is a free L site, with analogous definitions for ρ_s, ω_s, and Ω_s. Each of these quantities can be calculated using the methods developed by Flory and Stockmayer for their theory of gelation discussed in Chapter 11, or by percolation on Bethe lattices. If closed loops are forbidden, then Goldberg (1952) showed that for

$$p = p_g = \frac{1}{Z - 1}, \tag{14.4}$$

there is a finite probability that an infinitely large aggregate exists. This is the same as the site (or bond) percolation threshold of a Bethe lattice of coordination number Z (2.1).

The problem with this formulation is that the formation of closed loops is forbidden. Thus, the results that were obtained represent the mean field limit of percolation, i.e., a system whose spatial dimension is six or higher. In reality, antigen–antibody aggregation is a three-dimensional phenomenon, and therefore we have to use percolation on finite-dimensional lattices, and in particular three-dimensional ones, for explaining antigen–antibody aggregation phenomena. But to the best of my knowledge, this has not been attempted yet.

14.2 Network formation on lymphocyte membranes

According to modern versions of the clonal selection theory, the cells of the immune systems arise from division and differentiation of the stem cells in the bone marrow (Burnet 1959). Those cells that are potentially able to secrete antibodies are know as B-lymphocytes. These lymphocyte cells insert a homogeneous set of antibody-like receptor molecules on their membrane surface where they are used for recognition of complementary antigens. The population of B-cells is itself heterogeneous, i.e., different cells may have different types of membrane-bound receptor molecules. Under appropriate conditions, binding of complementary antigen to the receptors on a B-cell will activate that cell to differentiate and/or proliferate into antibody secreting cells. Moreover, the antibody secreted by the progeny of a particular B-cell is assumed to have specificity for the antigen, identical to that of the receptor molecules on the B-cell progenitor. Therefore, the role of antigen is cellular selection and amplification and its mission is mediated by interaction with cellular receptors.

The process of proliferation and secreting large amounts of antibody is preceded by cross-linking of antigens to the receptors until a macroscopically large patch is first formed on the surface. This patch or network is the two-dimensional analogue of three-dimensional aggregation, discussed in the preceding section, a process called *network* or *lattice* formation. Antigen-stimulated lymphocyte lattice formation may play a key role in triggering immunocompetent cells. For immunogens with repeating arrays of antigenic determinants, e.g., polysaccharide flagellin, triggering is possible without the aid of helper cells, while other immunogens require the presence of T-cells or their products. It has been suggested that the T-cell requirement results from having to present the immunogen to the B-cells in an aggregated or network form, thus increasing its valence.

The relationship between network formation and biological activity is not completely clear yet. It is presumed that lattice formation provides some sort of stimulatory signal, but this may not be enough to trigger a cell. In any event, it is not unreasonable to assume the triggering to be a function of the strength of antibody–receptor interaction. Delisi and Perelson (1976) assumed that lattice formation is the two-dimensional analogue of the three-dimensional precipitation reaction discussed above. They treated the problem in a similar way. To simplify the problem they assumed that the effect of intramolecular reactions is unimportant. A well-understood characteristic of immune response to antigens is an increase in antigen–antibody affinity with time after immunization. The implication is that some aspects of the antigen–receptor interaction is equilibrium-controlled. However, since antigens bound by more than one receptor dissociate very slowly, such binding may be effectively irreversible on the time scale of lattice formation. Delisi and Perelson (1976) derived several analytical results for the problem, using (implicitly) percolation on Bethe lattices (by a method similar to that of Fisher and Essam (1961), including the cluster size distribution, the critical behavior of the system near the point at which a large lattice is formed, i.e., near the percolation threshold, and membrane transport. Their results cannot be directly compared with any experimental information, as the results with a Bethe lattice are certainly not relevant to a two-dimensional system. But their work definitely indicates the relevance of percolation to such biological phenomena.

14.3 Percolation aspects of immunological systems[1]

While immunology is an old research field, applications of percolation concepts to it are quite new, and some of our discussion here may not be upheld by research in the years to come.

The deadly disease of smallpox has been eradicated from the earth for many years, but against AIDS (acquired immunodefficiency syndrome) no cure has been found at the time of writing (1993). These two examples are perhaps the extreme cases of success, and the lack of it, in applications of immunology. So how does the immune system work? If we get the flu in one winter, we will not normally get the same sickness shortly thereafter. Our body's white blood cells or lymphocytes produce antibodies or other cells which are able to neutralize the foreign virus or other antigen. These antibodies fit the antigens like a lock fits a key. Apart from small inaccuracies, one type of antigen fits only specific antibodies, and vice versa. If we get a new virus, some antibodies accidentally (through mutations) may fit this antigen, our immune system notices this fit, produces more of these specific antibodies, and in this way combats the disease. When we get healthy again, a few very long-lived memory cells survive and allow a quicker response of our body once the same virus returns. Vaccination produces a controlled amount of sickness so that antibodies and memory cells are formed for this specific disease. The human immunodeficiency virus HIV seems to escape destruction by the immune system, and instead destroys slowly the T4 white blood cells, which are crucial as "helper" cells for the functioning of the immune response.

According to the now widely believed ideas of Nobel laureate Niels Kai Jerne (1974), antibodies can be treated in turn as antigens by the immune system, which then produces antiantibodies, antiantiantibodies, etc. Moreover, the inaccuracies of the lock-and-key relation between the various types of molecules let the same antibody fit slightly different antigens, and the same antigen may be neutralized by slightly different antibodies. In this way, all possible antibody shapes are connected, directly or indirectly, by antibody–antigen relations in a Jerne network spanning the whole shape space.

What does this have to do with percolation? When we are ill, we take a medicine specific for that illness; not all possible types of medicines have been invented yet. Similarly, our immune system would destroy us if a flu infection could trigger all possible shapes of antibodies. Thus, the immune response should be very specific, i.e., the antibodies triggered by one specific antigen should form a finite cluster in the topological network of possible shapes. They should not percolate throughout this shape space. In other words:

---

[1] This section was written by Dietrich Stauffer.

WARNING: Immunology has determined that percolation is dangerous to your health!

de Boer and coworkers (1989, 1992) and Stewart and Varela (1991) have therefore investigated the conditions under which the immune response remains limited to a few types of antibodies, once a foreign antigen enters our body, i.e., the conditions under which percolation does not occur. Inherent in percolation theory, as presented in this book, is the clear distinction between occupied and empty sites. Thus, percolation is the easiest to apply to immunology if antibody molecules are either there or not there, without a more quantitative and more realistic distinction between such molecules according to their concentration. Indeed, such "cellular automata" approximations (i.e., discrete systems in which each site can be in one of only two states, sick or healthy, and nothing in between) have already been developed in immunology by Kaufman *et al.* (1985) before percolation aspects were investigated. The combination of percolation and cellular automata methods then allowed computer simulation of five- to ten-dimensional shape spaces (Stauffer and Sahimi 1993a, b), as required for natural immune systems according to Perelson and Oster (1979).

Such studies of localization versus percolation can be done in two ways. We may look at the immune response, if no element of the immune system has yet been triggered. Then our clusters are simply the sets of activated antibody types connected by Jerne network bonds (Neumann and Weisbuch 1992). Alternatively, we may look at an immune system which has already evolved into some stationary equilibrium of present and absent antibody types and check for the *changes* made to this dynamic equilibrium by one specific type of antigen (Stauffer and Weisbuch 1992, Stauffer and Sahimi 1993a, b). This is analogous to the chaos studies in dynamical system and has been used before in genetics (Kauffman 1969) and Ising magnets (Creutz 1986); physicists often call it "damage spreading" whereas pattern recognition neural networks experts talk about the "Hamming distance". Damage is the number of sites which differ in their spin or other characteristic values in a site-by-site comparison of two lattices, if the initial configurations of the two lattices differ by only one site. Hamming distance is used in neural networks to describe the number of pixels which differ in a comparison of two pictures. In particular, if one picture is the original aim and the other picture is an attempt to restore that original picture from a blurred or noisy version of it. Thus, in this sense damage spreading and the Hamming distance are the same.

At present it seems that for the immunological models investigated, the question of whether the immune response remains localized or percolates through the whole system depends on the parameters we choose, hardly a surprise to anyone familiar with the concept of a percolation threshold.

14.4 Protonic percolation conductivity in biological materials

Water–macromolecule interactions can influence folding, enzymatic activity, and other properties of globular proteins. One way of studying various properties of such biological systems is through measuring their hydration-dependent dielectric losses and electrical conductivity in the hydration range critical for the onset of enzymatic function. Behi *et al.* (1982) showed that hydrated protein powders exhibit dielectric dispersion at three different frequencies. The first one, $\Delta\varepsilon_1$, occurs at frequencies near 1 Hz, with a corresponding change in the DC conductivity of the system, which has been attributed to an isotope effect. The second and third dispersions, $\Delta\varepsilon_2$ and $\Delta\varepsilon_3$, occur at much higher frequencies, 10^5 and 10^{10} Hz, respectively. Bone *et al.* (1981) and Behi *et al.* (1982) proposed that these two dielectric dispersions are due to Debye relaxation of water bound on the surface of the macromolecule.

Careri *et al.* (1985) made careful measurements of dielectric losses for lysozyme powders of various hydration level in the frequency range 10 kHz to 10 MHz. The powders were prepared from Worthington thrice crystallized and salt-free proteins. The dielectric permittivity of the protein is not measured directly because the sample is only one part of the composite condenser that consists of a layer of the powder included between layers of dry air and glass. However, in the frequency of interest to Careri *et al.* glass does not display appreciable dielectric loss. Thus, the system is essentially a capacitor consisting of two layers of similar thickness, one of which has a vanishingly small conductivity and a dielectric constant close to that of the vacuum. The dielectic relaxation time t_d is then predicted to be

$$t_d = \frac{\varepsilon_0}{g_e}\sqrt{1+\varepsilon}, \tag{14.5}$$

where ε_0 is the vacuum permittivity, ε is the relative dielectric constant, and g_e is the conductivity of the hydrated protein. It was confirmed that the system is insensitive to slight changes in thickness of the air or powder layers. In the frequency range of interest, the hydration dependence of ε is much weaker than that of g_e. Thus, a plot of t_d versus h, the hydration level, is essentially equivalent to a plot of g_e versus h.

Figure 14.1 displays the resulting conductivity as a function of the hydration level h. In order to remove the nearly negligible contribution of nonpercolative processes and systematic errors in the evaluation of capacitor geometry to the total conductivity, the value of g_e at the percolation threshold is subtracted from g_e at any other hydration level h. It was established that protonic conduction is the dominating contribution to the dielectric relaxation $\delta\varepsilon_2$ in the frequency range in which the experiments were carried out. This relaxation is attributed to proton displacements on a

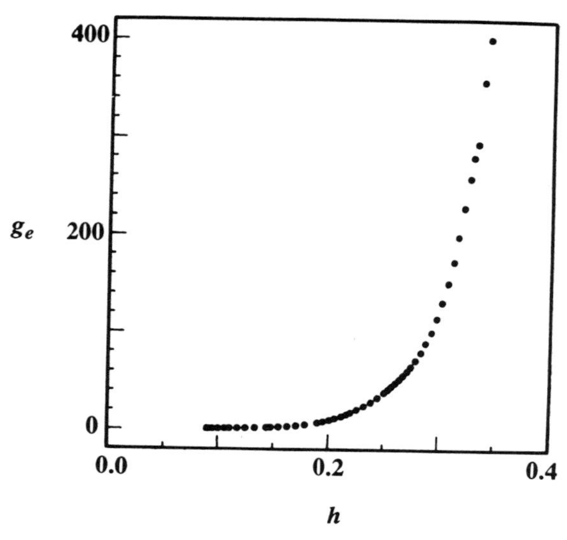

Figure 14.1 *The dependence of the effective conductivity g_e on the hydration level h in lysozyme powder for pH = 7, at 301 K. g_e has been normalized with the conductivity of the dry sample (after Careri et al., 1988).*

single macromolecule. Since h is proportional to p, the usual occupation probability in percolation, we can write

$$g_e(h) - g_e(h_c) \sim (h - h_c)^{\mu} \tag{14.6}$$

where h_c is the hydration level at the percolation threshold. The critical exponent μ was evaluated by Careri *et al.* (1988) for three different samples, namely, native lysozyme hydrated with H_2O, native lysozyme hydrated with D_2O, and 1 : 1 complex with $(GlcNAc)_4$ hydrated with H_2O. The results are shown in Fig. 14.2. There are two distinct regions; a region where we obtain a critical exponent $\mu \simeq 1.3$ and a region where we obtain $\mu \simeq 2.0$. The first region is interpreted as being an essentially two-dimensional system, since the system is relatively thin. In the second region, at higher hydration levels, intermolecular water bridges are established, giving rise to a higher-dimensional system. Both exponents are in excellent agreement with those of two- and three-dimensional percolation (see Table 2.3).

What is the mechanism for this percolative conduction? Careri *et al.* (1988) argued that this percolation process consists of proton transfer along a thread of hydrogen-bonded water molecules adsorbed on the protein surface. In this interpretation, water molecules are equivalent to the conducting elements in percolating systems. The mean free path of the protons at or above h_c is the distance between the poles of the macromolecules set by the boundaries of the molecule. The local structure of the protein itself

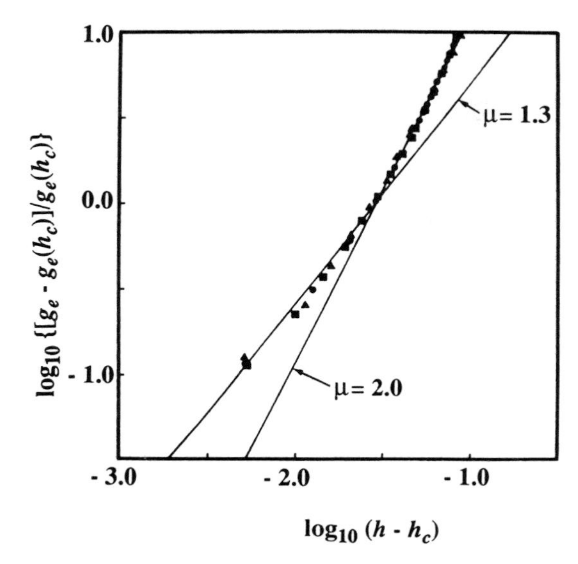

Figure 14.2 *Estimation of the critical exponent μ of the effective conductivity of lysozyme powders of pH = 7. Symbols denote the data for native lysozyme hydrated with H_2O and D_2O, and a sample of 1: complex of lysozyme with $(GlcNAc)_4$, hydrated with H_2O (after Careri et al. 1988).*

is not important, only the structure of the clusters made of water molecules acting as interconnected conducting elements is important.

These results have important biological consequences. The protein system used in these studies, with its surface sparsely covered by water or conducting elements, is similar to a protein membrane whose internal surface is sparsely populated, although the membrane itself is immersed in a solvent of near unit water activity. Percolation then suggests that membrane conduction is possible with channels only partly filled with conducting elements. Moreover, conduction can be turned off or on by adding or subtracting a few conducting elements, without changing the basic structure of the membrane. Finally, since percolation focusses on the effect of randomness on conduction, we can bypass the need for a high level of structure extending over the full thickness of the membrane or its hydrocarbon core.

Rupley *et al.* (1988) extended these studies to lightly hydrated purple membranes, more complex than the protein systems. The conduction paths may be predominantly within one of several regions of the membrane, such as the lipid surface, the lipid–protein interface, or entirely within the protein. The existence of such preferred paths may give a two-dimensional character to the conduction process in such membranes. Indeed, the measured critical exponent $\mu \simeq 1.23$ is close to that of two-dimensional percolation, although the structure of the system may be three-dimensional. Bruni *et al.* (1989) extended such studies to a dry tissue where a conductivity process will be integrated in complex, living systems. They used tissues of maize seeds, where water-induced effects lead to the onset of integrated metabolism, and thus to germination. The protonic nature of the conduction process in such tissues was established by deuterium substitution. The critical exponent for the conduction process was found to be $\mu \simeq 1.23$, close to that of two-dimensional percolation.

14.5 Conclusions

Percolation theory can explain certain aspects of the complex behavior of some biological systems. Moreover, percolation helps us to understand certain features of immunological systems, which are so crucial to the survival of living beings. Finally, the experimental verification of percolation conductivity scaling law in biological systems is not only a nice demonstration of the wide applicability and usefulness of such scaling laws, it also has practical implications for the structure of biological membranes.

References

Aladjem, R. and Palmiter, M.T., 1965, *J. Theor. Biol.* **8**, 8.
Asakura, S., Eguchi, G., and Lino, T., 1968, *J. Molec. Biol.* **35**, 327.
Behi, J., Bone, S., Morgan, H., and Pethig, R., 1982, *Int. J. Quantum Chem. Quantum Biol. Symp.* **9**, 367.
Bell, G.I., 1971, *J. Theor. Biol.* **33**, 339.
Bone, S., Eden, J., and Pethig, R., 1981, *Int. J. Quantum Chem. Quantum Biol. Symp.* **8**, 307.
Bruni, F., Careri, G., and Leopold, A.C., 1989, *Phys. Rev. A* **40**, 2803.
Burnet, 1959, *The Colonal Selection Theory*, Nashville, Vanderbilt University Press.
Careri, G., Giansanti, A., and Rupley, J.A., 1988, *Phys. Rev. A* **37**, 2703.
Careri, G., Geraci, M., Giansanti, A., and Rupley, J.A., 1985, *Proc. Natl. Acad. Sci. USA* **82**, 5342.
Creutz, M., 1986, *Ann. Phys. (NY)* **167**, 62.
de Boer, R.J. and Hogeweg, P., 1989, *J. Theor. Biol.* **139**, 17.
de Boer, R.J., Segel, L.A., and Perelson, A.S., 1992, *J. Theor. Biol.* **155**, 295.
Delisi, C., 1974, *J. Theor. Biol.* **45**, 555.
Delisi, C. and Perelson, A.S., 1976, *J. Theor. Biol.* **62**, 159.
Fisher, M.E. and Essam, J.W., 1961, *J. Math. Phys.* **2**, 609.
Goldberg, R.J., 1952, *J. Am. Chem. Soc.* **74**, 5715.
Hohn, T. and Hohn, B., 1970, *Adv. Virus. Res.* **16**, 43.
Jerne, N.K., 1974, *Ann. Immunol. (Inst. Pasteur)* **125C**, 373.
Kauffman, S.A., 1969, *J. Theor. Biol.* **22**, 437.
Kaufman, M., Urbain, J., and Thomas, R., 1985, *J. Theor. Biol.* **114**, 527.
Karnovsky, M.J., Unanue, E.R., and Leventhal, M., 1972, *J. Exp. Med.* **136**, 807.
Neumann, A.U. and Weisbuch, G., 1992, *Bull. Math. Biol.* **54**, 21, 699.
Poglazov, B.R., Borhsenius, S.N., Belozerska, N.A., and Belantseva, A., 1967, *Virology* **30**, 36.
Perelson, A.S. and Oster, G.F., 1979, *J. Theor. Biol.* **81**, 645.

Rupley, J.A., Siemankowski, L., Careri, G., and Bruni, F., 1988, *Proc. Natl. Acad. Sci. USA* **85**, 9022.
Stauffer, D. and Sahimi, M., 1993a, *Int. J. Mod. Phys. C* **4**, 401.
Stauffer, D. and Sahimi, M., 1993b, *J. Theor. Biol.*, in press.
Stauffer, D. and Weisbuch, G., 1992, *Physica* **A180**, 42.
Stewart, J. and Varela, F.J., 1991, *J. Theor. Biol.* **153**, 477.

Index